Flash Effect

Flash Effect

*Science and the Rhetorical Origins
of Cold War America*

David J. Tietge

Ohio University Press
Athens

Ohio University Press, Athens, Ohio 45701
© 2002 by David J. Tietge
Printed in the United States of America
All rights reserved

10 09 08 07 06 05 04 03 02   5 4 3 2 1

"Superman" from *The Carpentered Hen and Other Tame Creatures* by John Updike, copyright © 1982 by John Updike. Used by permission of Alfred A. Knopf, a division of Random House, Inc.

Illustrations "Blast Effect" and "Flash Effect" by Eric Mose from *Scientific American* used by permission of Eric H. Mose, Jr.

Illustration by Irving Geis from *Scientific American* used by permission of the Geis Archives Trust.

Library of Congress Cataloging-in-Publication Data

Tietge, David J., 1966–
    Flash effect : science and the rhetorical origins of Cold War America / David J. Tietge.
        p. cm.
    Includes bibliographical references and index.
    ISBN 0-8214-1433-X (acid-free paper)—ISBN 0-8214-1434-8 (pbk. : acid-free paper)
        1. Science—Social aspects—United States—History—20th century. 2. Technology—Social aspects—United States—History—20th century. 3. Religion and science—United States—History—20th century. 4. Rhetoric—Political aspects—United States—History—20th century. 5. Cold War—Social aspects—United States. 6. Popular culture—United States—History—20th century. 7. United States—Social conditions—1945- 8. United States—Politics and government—1945-1989. I. Title.

Q175.52.U5 T54 2002
303.48'3'0973—dc21   2002017543

Contents

Introduction

We do not here aim to discredit the accomplishments of science,
which are mainly converted into menaces by the inadequacies of
present political institutions.
—Kenneth Burke, *Permanence and Change*

Between 1903 and 1969, the United States transformed aviation from the Wright Brothers at Kitty Hawk to Apollo 11—a mere sixty-six years from sustained flight lasting only twelve seconds to putting human beings on the moon. In the span of a few decades, humanity has found cures for diseases that only a hundred years ago were fatal. We enjoy nearly instant gratification of the things we need and desire—provided we have the financial means—thanks in large part to better agriculture, transportation, and resource management and the ability to deliver goods quickly and efficiently. We have, to borrow a popular sentiment, made the world smaller by networking computers, phone lines, and satellites, making communication efficient and reliable. We enjoy the convenience of relatively cheap transportation to nearly anyplace we wish to go; distances that took weeks or months to traverse at the turn of the twentieth century now take only hours. We can predict weather more accurately, diagnose medical conditions earlier, and work with greater efficiency than ever before.

When we consider earlier technological history, we find no precedent for the speed with which we have progressed in the last century. Many factors may help explain this phenomenon, but none is more important than the connection between the unprecedented technological progress and the equally unprecedented destructiveness that the twentieth century produced. In the past one hundred years, we have witnessed two world wars that have taken the lives of tens of millions, we have endured many other major conflicts throughout the world, and we have glimpsed countless isolated skirmishes that may grace the headlines for a short time and then just as quickly disappear from our attention. This is more than coincidence. Technology and warfare have always been closely linked, but never more so than during a period when new sources of power, and new methods for harnessing it, were at the disposal of scientists, engineers, and military personnel.

Power, in this sense, carries a double meaning. Technological power means the discovery and control of new forces—electricity, steam, internal combustion, hydraulics, and . . . nuclear energy. Those who understand, develop, and control such power also gain power over those who do not—or, at least, they can turn it over to those already in a position of power so that they might use it to maintain or increase their influence. The military, of course, is the most obvious recipient of technological development, since it is through this agency that technological development is frequently endorsed, facilitated, and funded. Many modern conveniences, from microwave ovens to computers, found their initial use in (and were developed specifically for) military applications, and even those that did not were quickly given a military function. It should not be surprising, therefore, that military leaders and scientists have enjoyed a mutually beneficial, if at times reciprocally wary, coexistence.

Given the speed of our technological evolution, and the clear link between the military and the scientific advances that fostered that evolution, certain questions arise. Would we have advanced so rapidly had military tensions not existed to help create a context (and an incentive) for the discovery of new technology? How vastly different would the technology that we now possess be had not such tensions existed? How might our overall attitudes regarding technology (and society in general) have changed given more peaceful conditions for scientific discovery? Would we practice science differently, and would we view science with the same reverence that we now do had history unfolded along another, more benign, path? Finally, what is the relationship between national attitudes regarding science and the technological might a nation holds?

As important as these questions are, none is easy to answer. One problem is the sheer intricacy of these connections. It is difficult to assess, for example, whether science fostered greater aggression during the twentieth century because of the new technology available, or whether greater aggression pushed science to find technological solutions to military problems. And to assume that these are the only two possibilities would be to commit the fallacy of the false dilemma. It is quite possible that, with each new discovery, new and more malicious uses were hatched, and, as a result, many scientists were quick to jump on the bandwagon of national endorsement for their projects. Of course, each technological development has its own story to tell, and it would be unproductive to assert a simple causal connection between military desires and technological advancement, even if, in many cases, this happened to be the trend.

But we *can* say with certainty that throughout the course of World War II, American science and the military had established a reciprocal arrangement, one that had a common ideological cause—to win the war. To that end, the scientists of the Allied forces were charged with the development of new technology, from innovative aircraft design to the invention of radar and sonar to the creation of more destructive weaponry, which culminated in the design of an atomic bomb. All of these developments enjoyed widespread popular support in the United States, and the decision to drop the bombs on Hiroshima and Nagasaki was considered by most laypersons as a necessary evil, one that, while terrible in its implications, meant a speedier end to the Pacific conflict and the safeguarding of thousands of American lives.

But once the bombs had been dropped, and the Japanese surrendered, the fallout was more than radioactive; it was ideological. Science had unleashed an awesome power capable of leveling not just buildings or factories, but also whole cities. The implications were enormous. The United States was now sole possessor of a device that would solidify its domination of the globe in the early postwar years. We had, in the words of Harry Truman, "taken the biggest scientific gamble in history, and won." Far from turning people away from science, the Hiroshima event reinforced an ideological faith in the power of science and technology; it confirmed for Americans an idea that had been evolving since the Enlightenment: that science would help us overcome any obstacle, any conflict, any problem. Rational thinking and the scientific method were the panacea for all our social, political, economic, and diplomatic ills, and America, by beating its enemies to the development of the atomic bomb, had proved that it was the master of science. Science was not only a method and a practice; it was a commodity, a calling, and even a belief system.

Science functioned for Americans as a representation of their role in the world. Technology, and, in particular, nuclear weaponry, was a physical manifestation of that role. Reverence toward science made it almost a secular religion, and people went into scientific fields in unprecedented numbers, thanks in part to the GI bill, a piece of American legislation designed (among other things) to draw veterans to educational opportunities in science and technology. It was important, American policy makers reasoned, to maintain our foothold in scientific enterprises in order to preserve our technological advantage. This meant that we would need more people trained in scientific and technological fields. America had invested much in winning the war in Europe and the Pacific, at great cost, sacrifice, and risk; it had an economic and ideological stake in seeing that investment flourish.

But politicians and military leaders alike knew that our advantage would not last, and soon our former ally became our ideological enemy. It was not long before the Soviet Union, the other great victor of World War II, would also secure atomic weaponry for itself, and their cost, sacrifice, and risk during World War II made U.S. losses look like a stubbed toe. This meant that their investment in preserving their place in global politics was even greater than ours, and the United States needed little rationale to sell the importance of increased scientific and technological development to the American public. And while it is true that anxiety regarding nuclear war was at its most acute during the early Cold War, the only solution that seemed to be realistically entertained was increased research into, and development of, scientific knowledge convergently manifested in nuclear weapons.

During this time of cold conflict, science was, in the words of the late Carl Sagan, the "light in a demon-haunted world." It was through science that we could keep the Soviet threat at bay, it was through science that we would show our moral and ideological superiority, it was through science that we could rally as a nation, and it was through science that we would demonstrate our grace as God's chosen people. Science was not merely an objective methodology for uncovering a truth that existed in the physical world; it was a paradigm for our principles as a nation. It was a "god-term," to adopt the parlance of Kenneth Burke, which gave us purpose and direction. It was a touchstone for our ethics, and it was used as a *rhetorical* tool for helping us supplement a worthy, and even preordained, cause. The early Cold War era demonstrates the split that science has undergone since it has been gradually introduced, albeit incompletely, to the public way of thinking. As a rhetorical agent, the language of science was (and is) used for political and social change, the incorporation of new policy, and the maintenance of a status quo that has been established by the advocates of the scientific enterprise. I should mention that these advocates are, by and large, not scientists themselves (there are, however, some important exceptions). Rather, they are those who find a rhetorical advantage in adopting the language of science for the circulation of ideas central to their agenda, usually in a political or corporate context.

Science, considered from this vantage point, has a linguistic dimension that is usually overlooked by practitioners and advocates alike: science is used rhetorically. In the words of Tim Crusius: "Science fills an important cultural need, which it also helped to create. Modernity can be characterized as the struggle to cope with an increasingly diverse and decentered society. Such a society necessarily becomes more and more muddled about how to interpret itself, less and less certain about and sensitive to the subtler dimensions of meaning so dependent on *sensus communis*" (75). It functions, that is, as a surrogate for a richer rhetorical tradition, one that relied on "connotation, tonality, and gesture," replacing it with a system of discourse that is ostensibly more neutral and impartial, based on the positivistic assumption that "Truth" is a stable condition just waiting to be discovered through the empirical method (75). Thus science is a methodology, but it is also inherited by the nonscientist as an ideology of "facts," such that "when people say to us, 'These are the facts,' they mean, 'You should think this way' or 'Do this,' choose one course of action with its associated values rather than another" (77). How far does society go to privilege science over other ordering systems such as religion or the "humanities"? If the Cold War is any indication of science's status during a time of diplomatic crisis, very far indeed.

Following the appearance of Thomas S. Kuhn's *The Structure of Scientific Revolutions* in 1962, interest in and questions about the nature of scientific paradigms emerged as scholars challenged science's unsullied reputation as an objective, infallible epistemology. *Paradigm* is a term that Kuhn popularized, and one that has since been a seemingly indispensable component of subsequent literary and critical theory. Though some consider it jargonistic, the impetus behind the concept is far from trivial, and, at the time Kuhn wrote *Structure,* it was a pioneering idea. Kuhn aroused considerable anxiety among scientists, as early responses to his ideas indicate. As a former physicist, Kuhn was an inside member of the scientific community, so his clarion call as an advocate of scrutinizing scientific discourse was perhaps not as unexpected as we might think.

In the wake of World War II, Kuhn and other scholars questioned not only science's ethical basis but also its methodology and its production in a world that had increasingly trusted science as the lighthouse of knowledge. Kuhn had made the aggressively obvious (but as yet virtually untapped) observation that scientific discoveries were not the result of simple methodological disclosure—discoveries that every rational person would view as indisputable—but in fact new theories often resulted from a much more complicated social and political process in which questions of validity disguised more prejudicial professional, social, personal, and, most importantly, political interests. Kuhn's work initiated particular interest in the interior rhetoric of scientific discourse—that is, the rhetorical nature of communication within the scientific community itself. Much less attention has been given to how scientific knowledge is filtered down to the general public.

One way to discern how science has a rhetorical impact on the public—through media vehicles, distorted information, and outright propaganda—would be to examine the public dissemination and consumption of scientific lore in a period of great social and epistemological change. One such period was the Cold War era immediately following World War II. During this early Cold War era (circa 1947-60), increased armament buildup, experimentation with nuclear weapons, and even the rudiments of a space program were rationalized as scientific achievements marking U.S. superiority in technological progress. To what extent, one might ask, was the necessity of scientific advancement real (that is, responding to a bona fide Soviet threat against the United States), and to what extent was the threat rhetorically perpetuated so that scientific and technological expansion could continue unencumbered as it had during World War II? Were the policy makers of the early Cold War attempting to maintain the legislative freedom they had enjoyed during the war? And, most important, to what extent did the language of science itself legitimate the Cold

War conflict in the name of national defense to a public that must endorse the doctrines science forwarded?

Kenneth Burke and Michel Foucault provide useful perspectives from which to address such questions, but their work in the area of public conceptions of science is usually presented through a derivative exegesis that elucidates their own scholarly projects. Burke, for example, seems primarily interested in using scientific fields such as psychoanalysis to make exemplary comment on the rhetorical nature of any established system of knowledge, especially the way it relates to literary theory (see both *Permanence and Change* and *A Rhetoric of Motives*). Burke was, however, deeply interested in the language of the Cold War—a period of American history during which he was at his most theoretically prolific. The Cold War, I think, had a large impact on the direction of his rhetorical theory; so much so that some analysis of this rhetorically turbulent period in relation to Burke's rhetorical theory will be touched on and referred to throughout this book.

Foucault treats science for its knowledge-power properties to show how it is often used as a discourse of control (see *The Order of Things, Madness and Civilization, The Birth of the Clinic,* and, to some lesser extent, *The History of Sexuality* and *Discipline and Punish*). Foucault, of course, was more interested in science as it became an emerging paradigm for the seventeenth and eighteenth centuries in Europe, especially France. Still, his theoretical basis is sound, and the language he developed from it may help elucidate the phenomenon of the Cold War in the United States. While both of these thinkers provide keen insight into the veiled linguistic properties of scientific discourse, their deliberately incomplete treatment of science as a central Western epistemological mind-set invites further study in more particularized areas.

Other scholars, such as Alan Gross, Lawrence Prelli, Trevor Melia, and Paul Feyerabend, have examined (and in some cases dedicated their careers to) the rhetorical constructs that are a feature of scientific discourse, but they rarely discuss the uniquely public nature of science as a mode of persuasion. Scholars in the field of rhetoric and composition, such as John Nelson, Allan Megill, and Deirdre McCloskey, in their collection of essays *The Rhetoric of the Human Sciences,* primarily examine the relatively apparent rhetorical idiosyncrasies of sociology, psychology, political science, and economics, and these primarily on an interior level (that is, how scientists of these disciplines rhetorically address others in the same professional community). Even John Campbell, who has limited his work to a single text, Darwin's *Origin of Species,* investigates the rhetoric of science in a restricted way, portraying Darwin as a rhetorician *and* a scientist.

It is important, then, to redirect the discussion in order to understand how the language of science significantly affects the Western, and especially the American, collective lay consciousness during the years following World War II. It is particularly important to examine the nature of the scientific episteme as one that has, in its function as a kind of secular religion, actually taken on many pious features in its winding path to the public mind. In terms of its rhetorical function, science has its "priests"— the elite group of scientists in the upper echelon of the scientific hierarchy who function as the guardians of knowledge and the protectors of methodological, theoretical, and conceptual dogma. Science has its parishioners: those professors, instructors, and teachers who further the word of good science in hopes of converting, or at least directing, other potential members of the scientific community. And if we accept the premise that science is a form of secular religion, then media vehicles such as television, radio, and the Internet are the technological pulpits from which popular science is preached. Science, then, becomes the new opiate of the masses in that it retains the qualities of a religious hierarchy, but it also functions, from a public point of view, as a doctrine of Truth—it has the capacity to salvage the waning soul of humanity because it can capitalize on the rational, pragmatic certainty that we crave to provide order. This is not to say that science's contribution to knowledge, especially technological knowledge, has not been great or influential; on the contrary, that science has managed to produce tangible results and predictive accuracy like no other epistemic methodology only speaks to its mythos as the great purveyor of knowledge. Thus it is especially important to study the perceptions of science as they fall on the doorstep of the lay public and the ways in which the liaisons between scientific knowledge and public understanding cultivate such perceptions.

In terms of what Kenneth Burke would call an ideology, then, there are distinctive features that need examination if we are to see the connection between religion, science, and how these seemingly disparate institutions share much when they are encountered by, even doled out to, the general public. Burke himself puts it this way:

> There is perhaps no essential difference between religious and scientific "foundations," however, as religion is probably the outgrowth of magic, itself a "science" based on theories of causation which were subsequently modified or discredited. Magic, religion, and science are alike in that they foster a body of thought concerning the nature of the universe and man's relation to it. All three offer possibilities . . . in so far as they tend to make some beliefs prevalent and stable. (*Counter-Statement* 163)

Religion and science share one important and frequently overlooked feature—they both order our world. From a rhetorical perspective, this gives science an associative dynamic that is powerful indeed. Science has enjoyed the privileged status in its capacity to order in the last couple of centuries because it deals empirically with measurable, quantifiable data; religion, of course, requires faith in something directly antithetical to this stance, and its momentum as an ordering presence in modern and postmodern society has diminished as a result of the conflict between these two competing ideologies. Rhetorically, however, the process is very nearly the same; while science is concerned with gathering information to produce knowledge, its reason for doing this can be considered fundamentally religious—it wishes to secure man's place in, and understanding of, the universe. Even if science is a replacement of religion, as is popularly held, it retains residual elements of the Christian ordering system in the attitudes of the general public; and while we may not consciously realize it, such a connection buttresses science's authority, giving it a rhetorical strength that pure science alone could not achieve.

This notion is unequivocal because it is useful for understanding how the radical nature of Cold War rhetoric could enjoy such an acceptable place in the mind-set of the post–World War II public. Science had, after all, in the minds of most Americans, essentially won the war in the Pacific. Our scientific and technological superiority had allowed us to stave off the aggression of the Japanese (though the Japanese were essentially beaten long before the dropping of the atomic bombs on Hiroshima and Nagasaki) and had also proved our resolve against the Germans by beating them to the development of the atomic bomb, a weapon that, Americans were rightly convinced, would have been used for world conquest in the hands of such a malevolent adversary. Through a kind of technological manifest destiny, we had been proved righteous through both our ability to produce the weapon and through our prudence in using it. This validation would set a precedent for the Cold War development of nuclear weapons and would convince a proud American public of our right (even obligation) to do so. Such weapons became both a physical and rhetorical arsenal for keeping the world safe for Democracy.

Since advocates and practitioners of science tend to think of their methods and goals as self-legitimating truths, truths that avoid the intellectual messiness of philosophy, religion, or even rhetoric, it is exactly this image of self-legitimation that is cultivated for consumption by the lay public. Most scientists deeply involved in their own projects would deny any concern with public perception of their endeavors. But, ironically, only those distributors of knowledge with privileged public status can enjoy such a cavalier attitude. The maintenance of this public image requires attention to the rhetorical concerns that exist between the creators of knowledge and a public who would, ostensibly, stand to benefit from it most. The result is a deliberately nurtured mystification of what scientists do when they conduct science.

This book examines the rhetorical relationship between science and a public that views it as an intellectually untouchable, mystical process of obtaining knowledge and truth and projecting ideological rectitude during the linguistically manufactured crisis of the Cold War. Much can be gleaned from close examination of the texts used to buttress the ideologies of post-World War II American society: propaganda, public service announcements, training films, news media and their adjacent features (political cartoons and editorials), and even the popular literature and film of the era. This project provides a foundational overview of the theory necessary for the analysis of rhetoric of science, discusses the historical nature of the scientific mystification process, and examines the scientific aspects of Cold War rhetoric as a representative example of how the rhetoric of science is used for politically, militarily, and socially propagandistic ends. Detailed analysis of the machines of war themselves give way to close attention to the socialization of military and scientific developments so that we might understand how the cultural and ideological entanglement that was the Cold War was kept under tenuous control by those in power. The idea is to contextualize the rhetoric of science in one pivotal period of American history, namely, the decade and a half following World War II.

   This can be accomplished, I think, by showing how a "cold" war needed to be developed as a replacement of the real war America had victoriously emerged from. That is, while it was much easier to justify the development of a superweapon such as the atom bomb during a time of obvious crisis, it was much more difficult to continue the development of these weapons after peace had been declared. The United States needed to identify (and had an interest in maintaining) an ideological enemy—in the form of the Soviet Union—to rationalize continued scientific research in general and nuclear arms development in particular. It needed, in short, a way to nourish and hold in readiness the war machine during a time when no outright hostilities existed. But this was by no means the only source of scientific motivation. The general welfare of the nation was at stake, and scientists, politicians, academics, and the public at large needed reassurance that the American way of life was strong and secure. To accomplish this, the United States needed a predominantly rhetorical weapon to garner public support for continued experimentation with its military one. The rhetorical weapon took the form of the anticommunist propaganda of the 1950s Red Scare and extensively used an unexpected and subtle miscellany of ideological references in the popular press to fuel and justify the scientific rhetoric that would ensure America's ethical and political foundation.

Chapter 1

Theoretical Perspectives on the Rhetoric of Science

Those who study the rhetoric of science have two main areas of interest: the disciplinary rhetoric of scientific discourse, which involves rhetorical analysis of what scientists do when they communicate with one another, and extradisciplinary influences, or how cultural, political, and social factors affect and are affected by the language of scientific discourse. When studying the disciplinary rhetoric of science, one might examine scientific studies published for professionals in a given scientific field, manuscripts that document areas of conflict between scientists themselves, or experiments that are published for the purpose of gaining funds and support for scientific projects. When studying science's extradisciplinary (or external) influences, two distinct subcategories are of interest: how scientific information is encapsulated and disseminated for understanding by a broad general audience and how extraneous factors, such as governmental intervention and contracting (which is often dictated by public support), affect what scientists can practically accomplish.

It is important to examine scientific discourse in its functionary processes as a form of language rather than as a form of mere methodology.[1] Burke's essay "The Four Master Tropes" in *A Grammar of Motives,* for example, reveals that the shift in our ordering apparatus is one that steps from a poetic realism (roughly speaking, a view of the world that relies on literary and rhetorical language—history, tradition, interpretation—to order our experiences) to a scientific realism (a view of the world that relies on empirical, observable data to order our experiences).[2] Unlike the earlier theory of poetic realism, science is not concerned with substance or being, but instead focuses on processes that ignore the motivations that drive these processes. Science is concerned, that is, only with discovering correlations, so that "when certain conditions are met, certain new conditions may be expected to follow" (Burke, *Grammar of Motives* 504-5). What we often fail to notice, however, is that science itself is driven by other linguistic contingencies that operate on and through it in its evolution as an accepted model: "And it is equally true that the discovery of correlations [in science] has been guided by ideational forms developed through theology and governmental law. Such 'impurities' will always be detectable *behind* science as the act of given scientists; but science *qua* science is abstracted from them" (505).

In its concern for correlation (for example, processes), science, then, attempts to abstract itself from linguistic forms it must necessarily use as soon as it moves into the social realm (the realm that concerns us most in this study). Mere correlation thus becomes inadequate because in its conversion from nonsymbolic motion to symbolic action, scientific knowledge—for example, the orientations suggested by methodical inquiry—extrapolates meaning from processes that in themselves have no intrinsic value. The socialization of scientific knowledge and statements about its significance occur in and through the properties of language that make it a shared system of meaning: "Human relationships must be *substantial*, related to the copulative, the 'is' of 'being'" (505). It must, in short, rely on the four master tropes—metaphor, metonymy, synecdoche, and irony—that we all use to communicate ideas. For example, when a cartographer uses mathematical calculation to chart a map, she is engaging in metonymy (a trope that can be equated with reduction), since the map "drawn to scale is a reduction of the area charted" (503); when a chemist describes the elective nature of certain molecules uniting with others, he is invoking metaphor (which can be equated with perspective), since the ability to elect is an activity that is possible only in autonomous beings; when a physicist uses the myth of Sir Isaac Newton's revelation of gravity by describing how he was hit on the head by a falling apple, she is using, among other things, synecdoche (which can be equated with representation), since the apple represents only one minuscule affect of gravity, though it is designed to stand for the concept of gravitational pull in the entire universe; when a social scientist describes the unintended consequences of technology, he is employing irony (which can be equated with dialectic), as when car alarms are ignored because it is assumed they are tripped accidentally, thus decreasing the intended effect of theft deterrence.

Tropes such as these play a key role in scientific discourse because they are the stanchions of symbolic expression. It is unlikely that we could express very much in any social context without them, in which case science would affect only a handful of people who were able to divorce themselves entirely (assuming, for the sake of argument, that this is possible) from their original sociolinguistic exigency. It should also be noted that these tropes are rarely encountered in a pure form, but have a tendency to "shade into one another" (503), which speaks most to our comfort in using them for expressive purposes. Science is no exception, and science invoked in a social setting, in fact, *relies* on such tropes to get the message across, lest we all be forced to talk in a strictly mathematical language (and even this, I doubt, could be entirely void of at least elements of these same master tropes).

While the theoretical establishment of science as a language system is fairly easy to demonstrate, how it combines with other institutionalized language systems is a far more daunting prospect. I mentioned earlier that Burke sees no *essential* difference (as opposed, perhaps, to a *substantial* difference) in religious and scientific foundations, claiming that they both seek (along with magic) an understanding of the universe and the part that human beings play in it. Given the history of religious dominance, one might ask how it is that science has managed to completely usurp religion as the dominant ideology in such a short span of time. The answer is: it hasn't. Religious symbolicity still permeates scientific concepts on many levels, but particularly in the way science is presented to the laity that was once the same civic body held under the spell of religion—a case of scientific realism replacing poetic realism. On an internal level, scientists try to avoid using religious imagery or theological references when discussing their work, deferring to the restricted language of their own discipline. But when this same work is displayed to civilians, handed down to their spokespersons, or presented for persuasion to nonspecialists, religion, religious references, and religious imagery are frequently invoked to demonstrate the significance of a finding or to "make some beliefs prevalent and stable" (Burke, *Counter-Statement* 163). (Consider one famous quote by the pope of physics, Einstein himself, "God does not play dice with the universe.") In the case of the Cold War, religious imagery, symbolicity, and value systems were used as a means to defuse, justify, and stabilize the chronic situation science had helped create.

Raymond Williams provides a useful theoretical basis upon which to explain this unexpected cultural development. In his essay "Base and Superstructure in Marxist Cultural Theory," Williams explains that "superstructure" is the "main sense of a unitary 'area' within which all cultural and ideological activities could be placed" (379), but that the "base" is "the real social existence of men" (379). More important for our discussion, however, is the notion of "residual, dominant, and emergent cultures." Williams means by residual cultures "that some experiences, meaning, and values, which cannot be verified or cannot be expressed in terms of the dominant culture, are nevertheless lived and practiced on the basis of the residue—cultural as well as social—of some previous social formation" (384). Even more significant is the notion that, in the case of residual cultures, "[t]here is a real case of this in certain religious values, by contrast with the very evident incorporation of most religious meanings and values into the dominant system [in this case, science]" (384). In other words, religious values are so pervasive in the base culture that there is a residual effect on the superstructure of the dominant culture. Religious history, tradition, and standards are residual, according to this theory, even in the ostensibly secular culture of science.

The result is an emergent culture, one in which "new meanings and values, new practices, new significances and experiences, are continually being created" (385). The values and ideological motivations of science and religion become conflated, so that it is difficult to tell where the ideology guiding one motivation ends and the other begins. A scientist might actively seek to unlock the mysteries of atomic power because he is genuinely interested in the scientific ideal of understanding the natural world; he might also be motivated, however, by a religious ethic that dictates that he must fight oppression and, paradoxically, murder. It would not be unusual, therefore, for this scientist—or a political advocate who might act as a mouthpiece for him—to describe his motivations *and his findings* in an amalgamation of scientific and religious nomenclature.[3]

We shall see in the case of the Cold War that, when the motivation driving the dominant culture is perceived as dubious, rationalizing language from an alternative, emergent language is enlisted in order to ensure the stability of the superstructure.

Further Theoretical Foundations for the Rhetoric of Science

Since the rhetoric of science is relatively new as a field of study and discourse, it is necessary to provide some further discussion of the theoretical premises that drive both this study and, to some degree, the field as a whole. The arguments for inquiry into the rhetoric of science are numerous, but I defer primarily to one. Many humanistic scholars feel that science has been exempt from critical scrutiny (at least on a level of textual or linguistic analysis) because most scientists and advocates of scientific methodology have maintained that such analysis is irrelevant to a field that deals with hard facts and physical data. Scholars who recognize the legitimacy of critically scrutinizing science, however, contend that the effects of science on society are hegemonic;[4] therefore, attention to the operations of science as a social and methodological *ideology* is necessary to understand the social motivations underlying the popularity of this very successful body of knowledge. In short, why do Americans (in particular) hold science in such reverence as a means of attaining knowledge? While certain answers to this question are obvious—that is, science supplies relative certainty and order to an otherwise apparently chaotic universe—the motivations behind the social uses of science are far more complicated.

As I have mentioned, Kenneth Burke provides some of the most provocative observations for pursuing questions about the linguistic and rhetorical nature of scientific discourse, and it is from his rhetorical theory that much of the groundwork for this project has been laid. Some of the most basic rhetorical premises for Burke can be found in his 1950 book *A Rhetoric of Motives*, where he discusses the key notions of identification and consubstantiality. Identification is, simply put, a "joining of interests" (either genuine or believed) between two or more people that leads to a "sharing of substance" (20-21).[5] While people may identify and therefore share substance with one another, this does not imply that there is a loss of identity in the individual: "two persons may be identified in terms of some principle they share in common, an 'identification' that does not deny their distinctness" (21). For two people to identify with one another, then, makes them consubstantial, a doctrine that, "either explicit or implicit, may be necessary to any way of life. For substance, in the old philosophies, was an *act;* and a way of life is an *acting-together;* and in acting-together, men have common sensations, concepts, images, ideas, attitudes that make them *consubstantial*" (21).

These two ideas of identification and consubstantiality are basic to any modern rhetorical analysis, for it is through these that we begin to understand how ideological conflicts like the Cold War can escalate. The Cold War demonstrates in a superlative way "the possibilities of classification in its *partisan* aspects" and reflects "the ways in which individuals are at odds with one another, or become identified with groups more or less at odds with one another" (22). Perhaps the best description of a conflict like the Cold War (or any war, for that matter) can be summed up as follows:

> Why "at odds," you may ask, when the titular term is "identification"? Because, to begin with "identification" is, by the same token, though roundabout, to confront the implications of *division.* And so, in the end, men are brought to that most tragically ironic of all divisions, or conflicts, wherein millions of cooperative acts go into the preparation for one single destructive act. We refer to that ultimate disease of cooperation: war. (You will understand war much better if you think of it, not simply as strife come to a head, but rather as a disease, or perversion of communion. Modern war characteristically requires a myriad of constructive acts for each destructive one; before each culminating blast there must be a vast network of interlocking operations, directed communally.) (22)

One source for the establishment of this communion during the Cold War, I will show, was science. (And given that *A Rhetoric of Motives* was written in 1950, I have no doubt that nuclear war weighed heavily on Burke's mind when he wrote this passage).

Since the "disease of cooperation" that is war is perpetuated by the human species, we should note that Burke's "Definition of Man" provides suitable criteria, for his understanding of the human animal is distinct from other animals. There are five criteria for man (I will use the masculine gender designation only for convenience, since this is the language Burke uses, though in later references to this essay he changes the title to "Definition of Human"): (1) man is the symbol-using animal; (2) man is the inventor of the negative; (3) man is separated from his natural condition by instruments of his own making; (4) man is goaded by the spirit of hierarchy; and (5) man is rotten with perfection (Burke, *Language as Symbolic Action* 4–15).

The first condition is one of the most important, though perhaps one of the most obvious. To illustrate the importance of symbol systems in human activity, Burke likes to refer to birds as examples of species incapable of the same process. The most revealing is the example of the parent wren that, having difficulty coercing one of its children from the nest, used an act of genius to accomplish the task. By placing a mouthful of food just out of the reach of the baby wren, the parent forced the baby to lose balance enough so that the parent could easily throw it out of the nest. The significance of this is threefold, according to Burke: first, had the wren possessed a sophisticated enough symbol system, it could have communicated this technique to other birds who might never have discovered it on their own. Second, it is probable that this wren never used this method again because it does not have a symbol system necessary for the "concept-utilization" of the idea. Therefore, it was unable to cognitively categorize the success of the technique for future use. Third, and luckily for the bird, since the wren possessed no such symbol system, it was immune to the abuses of language to which all human beings are prone (4–5).

The issue of man as a symbol-using animal is of course relevant because we must remember that, as a result of such symbol-using, "overwhelmingly much of what we mean by reality has been built up for us through nothing but our symbol systems" (5). This observation comes as a great difficulty to us, especially to those with scientific leanings, because we tend to mistake the symbol-driven nature of a situation for that of a real condition for the simple reason that our symbol usage is so ubiquitous. It follows, therefore, that by its proximity, language is rarely thought of as anything but a true representation of the world. To study the illusions that language creates is too difficult for us—so often we blind ourselves to the idea that language orders our world—and we tend to favor the opposite assumption. In short, "however important to us is the tiny sliver of reality each of us has experienced firsthand, the whole overall 'picture' is but a construct of our symbol systems" (5). For Burke, the important question, therefore, is "[w]hich motives derive from man's animality, which from his symbolicity, and which from a combination of the two" (7).

The second clause, "man is the inventor of the negative," arises from the fact that no negatives exist inherently in nature except where humans have superimposed them. It could be as easily argued that no positives exist either, but the question of whether humanity sees the world as intrinsically positive or negative seems a moot point; in either case, it is a manufactured distinction. Nature exists independently from any such charged nomenclature. An earthquake is a natural, and regular, occurrence, but when we lay the template of symbolicity onto it, it becomes a disaster or a tragedy. This is only one function of the negative, however. To negate something does not necessarily mean to find the bad. It may simply refer to a manner of taking away. In either event, it is a "function peculiar to symbol systems" (9), because we can look at any given object and go on ad infinitum describing what it is not. Also, we can relegate to it the function of unfulfilled expectations, as when we want a particular event to occur and a different one does instead, we say that the expected situation did not happen. This becomes a tricky philosophical exercise because the event that did not occur is merely a function of our imagination, opposed entirely to the event that did occur. We can imagine the events that did not occur, but they are completely removed from any empirical reality (10).

The third criterion, "man is separated from his natural condition by instruments of his own making," is especially relevant to the rhetoric of science since it implies not only the development of language but also the development of technology. Burke describes the reliance on technology as a kind of second nature among humans and cites the example of a New York City blackout. In the description provided by the newspapers, one was given the impression that the blackout was some unnatural, perverse occurrence, despite the fact that darkness, of course, is the wholly natural condition at night (13). He also posits that this third clause was developed to "take care of those who would define man as the 'tool-using animal'" because there is such a close connection between the use of tools and the use of language (13-14). While other animals use tools, none use tools to create tools, and none use words about words (what Burke refers to in *The Rhetoric of Religion* as logology). The example Burke provides is that while an ape may use a stick to bat down a piece of fruit from a tree, this same animal would not think to hollow out the end of one stick with a knife in order to insert another, thus creating an extension. This is the second level of tools and language, the ability to apply one level to create another. Language, then, is a "species of action, symbolic action—and its nature is such that it can be used as a tool" (15). While it is not entirely clear what man's natural condition is, it is clear that he is sequestered from nature, and, in fact, goes to great measures to block it out and control it. One could philosophically argue that it is man's natural condition to devise and use tools and language, in which case he is not separated from his own natural condition, but only following it. Still, the implications are profound enough to see the close interrelationship between symbol systems and technology, and this is what will concern us here.

The next criterion is that "man is goaded by the spirit of hierarchy," or that he is "moved by a sense of order." Burke suggests that this fourth clause is the culmination of the other three, since the other three are required to impose order. He cites a number of literary references that illustrate his point (such as the works of Dante, Castiglione, and Kafka), all of which represent the human urge to construct a cogent pecking order that allows for social cohesion. Science may be one of the most acute manifestations of this need to create order where presumably none exists, for it is unswervingly predisposed not only to categorize but also to assume that the categorization it creates is a natural condition of the workings of the world (as in a discussion of the laws of physics). This is another sticky philosophical point, for while it is undeniably true that the law of gravity is constant, to describe it as a law presupposes that human description is the essence of the phenomenon, when in fact it is an attempt to show the phenomenon in human terms—terms that reflect the human need to organize.

For the final clause, that "man is rotten with perfection," Burke says he must "hurry to explain and justify this wry codicil":

> The principle of perfection is central to the nature of language as motive. The mere desire to name something by its "proper" name, or to speak a

> language in its distinctive ways is intrinsically "perfectionist." What is more
> "perfectionist" in essence than the impulse, when one is in dire need of some-
> thing, to so state this need that one in effect "defines" the situation. And even
> the poet who works out cunning ways of distorting language does so with
> perfectionist principles in mind, though his ideas of improvement involve
> recondite stylistic twists that may not disclose their true nature as judged by
> less perverse tests. (16)

Burke claims that a precondition of language as a symbol system has a
principle of perfection implicit in it, so that we naturally wish to use correct diction, grammar, and try desperately to focus upon the correct
words. Perfection, as an abstraction, is of course illusory; a perfect language system would always be clear and unfettered by misinterpreta-
tion, but these qualities, Burke shows, are exactly what make language at
once rich with meaning and ripe for abusive applications. Human beings, however, are "moved by the principle" of perfection so much that
we often delude ourselves into thinking of it as an obtainable goal.
Science uses specialized language that it sees as close to this ideal of perfection, being exonerated of various shades of meaning and therefore always having the same objective definition.

We might follow up here by examining one of Burke's earliest comments on science, one that captures the importance of critically examining the rhetorical nature of scientific endeavors: "The flourishing of
science has been so vigorous that we have not yet had time to make a
spiritual readjustment adequate to the changes in our resources of material and knowledge." Burke is here referring specifically to the artistic
"breach between form and subject-matter" that "is the result of scientific
criteria being unconsciously introduced into matters of purely aesthetic
judgment" (Burke, *Counter-Statement* 31). Burke's observation applies to a
trend in literary circles that attempts to judge literary art using a quasi-
scientific standard, a result, he implies, of our scientific knowledge speed-
ing forward faster than our literary conventions (or wisdom) are capable.
Though the suggestion is specific to literature and literary criticism, the
phenomenon is one that proliferates nearly all (post)modern activities
since science is, as a worldview, "unconsciously introduced" as the exem-
plar for nearly everything we do.

For example, it is no coincidence that ideological notions of efficiency and economy began to surface during the time of the Industrial Revolution, for it was during this time that interest in and conditions for mass production were made possible. Though such ideals may have been valued in earlier times to a lesser degree, the conditions that made large industry possible were improvements in machinery and technology that allowed for maximum utilization of minimum resources, a movement made workable largely by improvements in science.[6] Efficiency, as an ideological paragon, is still extolled today, mainly because it springs from the scientific emphasis on progress and potential. Burke adds that the ethic of efficiency, however, if taken to its ideological extreme, becomes quite *inefficient* in practice. If production is the single most important goal of an efficient industry, then overproduction (and by extension waste) is the likely outcome: "whereas overproduction could be the greatest reward of applied science, it has been, up to now, the most menacing condition our modern civilization has had to face" (31). Such a standard, if implemented thoroughly enough, becomes a touchstone not only for machinery and resources but also of human activity. Hence we get such idiosyncratic phrases as "human resources" to describe human beings in their role as producers of labor. The overarching point, therefore, is that if we routinely, and unconsciously, invoke the criteria of science (as the contemporary human being seems comfortable doing on micro and macro levels), it can have the opposite effect of our intentions: we can become decidedly inefficient in our relentless pursuit of efficiency (and this, of course, is an instance of the ironic/dialectic trope in science).

Such an observation is not intended to disqualify science as a mode of knowledge; rather, it speaks to the ease with which we have been inculcated by the powerful, even seductive, persuasive acumen of science as a panacea for problems ranging from quantum physics to literary analysis. Yet science does not end its influence with the well educated or the elite. It has, it seems, an even tighter hold on the public as a whole, partly because of its mystique as a problem solver, and partly because the education the common person receives in science is usually incomplete, providing just enough to whet the imagination and to encourage faith in scientific systems if not actual knowledge of them. Most scientists fully realize the limitations of science (whether they admit it or not); most citizens of a scientific society do not.

If I am to argue the relationship between science and society, one might ask, on what basis is that relationship founded? The relationship lies, on a daily level, with technology. How, then, are technology and science linked? The answer is clear: the former is the applied manifestation of the latter. Before machines of contemporary sophistication could be devised, the scientific understanding of physical, geometrical, and mathematical laws was necessary. Technology is the praxis of theoretical science, and as such, is the thing that is handed down to a capitalistic society, in its most extensive form, as the consumer good. Right away it should be noted that a symbiotic relationship between science and politics is established; in like manner, the relationship between science and society is but another short step. Therefore, on a level of straightforward physical cause and effect, science permeates our society on a daily basis.

Its influence, of course, does not end here. Since the effects of science (in the form of technology) are so visible, so omnipresent, it makes sense that by sheer exposure we should begin to take the fruits of science for granted. We are, in effect, conditioned to internalize the very thing that encompasses our lives so that it is no longer a conspicuous entity, but becomes invisible in its enduring visibility, a definitive feature of science's hegemonic status. We climb into our machines daily without the slightest conscious acknowledgment that we have become extensions of those machines, or, perhaps, that the machines are extensions of us. We produce, write, calculate, move, communicate, work, think, and even feel with the aid of machines, while at once ignoring and paying homage to the benefits they provide. We *expect* them to serve us, and when they do not, as in the times when the "computers are down," we become paralyzed. We are trained socially to survive only with the aid of gadgets, and we are lost when they are not at our disposal. If we try living for a week without electricity, we quickly see how dependent we are on our "modern conveniences."

While all of this seems painfully obvious, what are subtler are the social and psychological effects this dependence exacts on us. To even use terms like *social* and *psychological* presupposes scientific criteria in the description, which is why it is so difficult to back away from the effects of science and look at its influence using a different nomenclature. We are jaded in our marriage to technology, because the linguistic superstructure that governs its use and its acceptance makes it exceedingly difficult to think in any other way. We reach what Burke refers to as a "trained incapacity," a level of social acclimation that blinds us to alternate ways of thinking and living. Burke borrows this term from Veblen to describe a certain function of orientation, a word that Burke uses to explain an "altered response" that results in "the outward manifestation of a revised judgment" (*Permanence and Change* 5). We learn, that is, to discriminate between things desirable and things undesirable, things accepted and things unaccepted, simply by living and experiencing the world around us. One outcome of this survival technique is trained incapacity, or the phenomenon that results in how "one's very abilities can function as a blindness" (7). Burke uses the example of a trout that after having escaped certain death by chomping into some bait and emerging with only a rip in his jaw will thereafter pass up anything that resembles the "jaw-ripping food." Trained incapacity sets in when, having learned this distinction, the trout passes up real food because it reminds him of the jaw-ripping kind. Trained incapacity has, as a survival method, actually impeded a survival contingency: eating.

Another example of this, cited by Burke but used by Veblen, has connotations more appropriate to human society: "Veblen generally restricts the concept [of trained incapacity] to business men who, through long training in competitive finance, have so built their scheme of orientation about this kind of effort and ambition that they cannot see serious possibilities in any other system of production and distribution" (7). I would submit that the same phenomenon occurs not only with scientists but also, more important, with the recipients of scientific production, namely, society in general. Having been trained (both formally and informally) in the language of science, it is often difficult for us to "see serious possibilities in any other system." As a result, we frequently ignore or even hold in contempt systems that do not meet the rigid requirements of scientific analysis.

This does not imply that all members of society are formally trained in scientific inquiry: most are not. It implies, however, that the rigors of the scientific method serve as a gauge for more intuitive modes of thinking; if these methods fall short of loosely understood scientific criteria, they are usually dismissed as meaningless. We have been so entrenched in the process of rational thinking (a byproduct of the scientific hegemony), that considering other epistemic possibilities becomes much less likely, or, at least, far less acceptable as a means of communicating ideas. However, such an understanding of our thinking process is itself an ideal, and the impression we have of ourselves as strictly rational beings is, of course, an illusion. We employ irrational judgments all the time, but we usually make them acceptable by attempting to rationalize them through a veneer constructed largely by the accepted mode of presentation. If it sounds rational, in other words, many are easily persuaded that it is.

Burke says several things about this rhetorical apparatus. In relation to trained incapacity, for instance, he discusses how we often use the idea of escape pejoratively to describe someone trying to avoid reality, as in the critical sentence "that movie was nothing more than escapist fare," implying that the movie had no foot in the realm of real or actual events or possibilities and appealed only to those with a propensity toward escapism. It seems unusual, Burke says, for us to attach such negative associations to a word that is a natural reaction to an undesirable situation. Further, in its restricted usage, escape is not considered a concept that applies to everyone, but only to those subject to avoidance and unable to cope with reality. Rhetorically, the problem with this is that "[w]hile apparently defining a *trait of the person referred to,* the term hardly did more than convey the *attitude of the person making the reference.* It looked objective, as though the critic were saying, 'X is doing so-and-so'; but too often it became merely a way of saying, 'I personally do not like what X is doing'" (8). This is a successful rhetorical strategy because it places the burden of reason on the person accused of trying to escape, while in fact revealing the subjective conviction of the accuser.

The connection between rationalization and orientation underscores a similar pattern. Rationalization, Burke posits, stems from locution developed in the "special terminology of motives" of psychoanalysis. For example, if I were to explain an event using the specialized language of Christianity (e.g., my wife died because it was the will of God), the psychoanalyst might reply that this was a rationalization designed to ease the pain of my loss. Where the psychoanalyst's explanation is analysis, my explanation is rationalization. It is the result of a process of self-deception that allows me to ignore the realities of the situation and explain them away under the responsibility of a higher authority (18).[7] The mystery, Burke says, is how one can denounce the religious explanation as rationalization when it may be the only one I have at my disposal. In other words, my ignorance of psychoanalysis hardly makes me a suitable candidate for self-deception when I have no knowledge of the psychoanalytic motive required to elicit such a deception. This is like "accusing a savage of self-deception who, having never heard of Pasteur, attempted to cure his diseases without the orientation of bacteriology" (18).

Moreover, the example of the religiously oriented rationalizer may go far toward explaining how and why we perceive scientific discourse as managing to usurp, and even replace, the authority of religious discourse in the common person. If religious-oriented language is dismissed as rationalization by those in a more favored discourse community (such as psychoanalysis), it follows that the language of the supported discourse community would necessarily serve as a substitute for the discredited orientation. Further, the new orientation might even take on some of the features of the old in the mind of the person whose previous language system has been discounted, since the associations inherent in the prior rationalizations are merely given a new linguistic designation (an example of the religious/scientific emergent culture). Therefore, it might not surprise us that the language of science is often replete with the religious connotations of the discourse it has so successfully replaced.

Is psychoanalysis, in this example, a more enlightened language system than the religious language system? Does it represent the truth more purely or more accurately? In terms of the motives it describes and the acceptance it demands, it does not really matter; we *act* under the belief that it does represent our behavior more efficaciously, and we therefore accept its analysis as true. The reason for this, Burke explains, has to do with a modern society that creates a more ambiguous sense of duties and virtues than earlier periods of human society, a social condition that may impair the "serviceability of our orientation" (21). He suggests that "our duties may not serve their purposes so well as they once did. Thus we may no longer be sure of our duties, with the result that we may cease to be sure of our motives. We may then be more open to a new theory of motivations than we should be at a time when the ideas of duty were more accurately adjusted to the situation" (21).

In general, the theory provided so far relates to the Cold War worldview in this specific way: by cultivating the orientation that scientific advancement was a necessary deterrent to Soviet domination, the U.S. government (in its myriad capacities) fostered a belief that only through

superior technology and nuclear upsizing could we hope to prevent a nuclear war and remain a globally dominant nation. While the logic of such
a contention is noticeably skewed, the American public accepted it almost unconditionally. The only way such logic can be sold as a cogent reality is through the prolific use of the kind of rhetorical theory that Burke
provides: our ideology, while complex, basically assumed that American
intentions were noble and worth pursuing just as our way of life was exceptional and worth protecting. Our trained incapacity was that we believed in our ideological propriety so completely that we were unable to
see the dangers in such thinking. All of these rhetorical problems were
furthered under the dictates of science, since it is through this hegemonic
pursuit that the problems arose in the first place, and it has been an un-
canny consistency in American thinking to assume that scientific solutions are the only ones at our disposal for scientific problems. Further, in
such speeches as Dwight Eisenhower's "Atoms for Peace" address, there
exists the real concern that the civic anxiety inherent over the Cold War
situation was so acute that it required a reorientation of the public so that
it might see new ways atomic energy could be applied. If this reorientation did not occur, civil fear could have led to social chaos.

But, one might ask, what else do we have to work with? Such a question itself reveals the degree to which science dominates our way of
thinking, that our trained incapacity is so acute that we cannot bring
ourselves to envision other possibilities (whatever they may be). This
trained incapacity is, albeit, immensely difficult to overcome. In *The
Rhetoric of Religion,* Burke draws a distinction between what he terms
Dramatism and *Scientism.* An approach couched in the latter might pose
such an "essentially epistemological question as 'What do I see when I
look at this object?' or 'How do I see it?'" while a typical question for the
orientation of the former might be "'From what, through what, to what
does this particular form proceed?' or 'What goes with what in this structure of terms?'" (38–39).

The distinction, then, is in what it is, exactly, we are looking *at.* A Scientistic approach, Burke says, begins with the problem of knowledge
and perception while the Dramatistic approach is more concerned with
the problems of action or form; one is epistemological and the other ontological. Both methods have distinct advantages, but Dramatism, Burke
argues, is more inclusive:

> The second rough-and-ready opposition is: A stress upon "action," as against a stress upon "motion." "Motion" is a necessary category, but if we treat all problems of motivation in terms of "motion," we get reduction *praeter necessitatem*. Behavioristic terminologies of motives would reduce "action" to "motion," whereas Dramatism holds that action is a more inclusive realm, not capable of adequate description in terms of "motion" only. "Action" is to "motion" as "mind" is to "brain." (39)

Physicists are concerned with motion; philosophers are concerned with action. We can observe, Burke points out, motion without action, though we cannot consider action without motion. This, he says, makes a theory of Dramatism a more complete picture of human activity since "there will always be an order of 'motion' implied in the realm of action" (39). Applied Scientism falls short of broad understanding of human conduct because it observes only the motion without possessing the analytical apparatus for understanding the motive behind it.

If the study of human conduct (in the broadest sense) is a study of action and not merely of motion, the study of language as a uniquely human instrument would be the study of language as symbolic action. There are four main realms of this symbolicity: poetics, or the aesthetic use of language; rhetoric, or language used to "induce cooperation by persuasion and dissuasion"; and ethics, or the language through which "we express our characters" (Burke, *Language as Symbolic Action* 28). The primary feature of each realm, of course, is that language is a representative component of action rather than a necessary one: "That is, we are the kind of animal that approaches everything through modes of thought developed by the use of symbol systems; what we don't have names for, we at least think of as 'nameable'—and in this respect we differ categorically from animals whose relation to their environment eliminates this roundabout, midway stage" (28).

In "Terministic Screens," an essay from *Language as Symbolic Action*, Burke notes that the very act of definition is itself a Scientistic view of how words work, because it is assumed that definitive language can somehow be exempt from the symbolism that governs all other forms of language. Definition, in the Scientistic view, dictates how words are used properly, as when we consult a dictionary to find the correct meaning of a word. The erroneous assumption here is that the words contained in the dictionary are not subject to the same problems of definition that arise with the original word that required definition. The definition seems pure and untainted by the symbolism, rhetoric, imagery, and sundry associations that plague other words, when in fact it is synecdochal in its attempt to represent the range of meanings and contexts through a single definition. Metalanguage such as that found in lexical definition has, we think, achieved a neutral status and is a proper instrument for describing other words. Science attempts to take this a step further by devising strictly objective, restricted, and ostensibly neutral language to describe its operations.[8]

A Dramatistic view of language, however, "is exercised about the *suasive* nature of even the most unemotional scientific nomenclatures" (45). The complexity of this observation becomes clear when we discover Burke's implication: "[e]ven if any given terminology is a *reflection* of reality, by its very nature as a terminology it must be a *selection* of reality; and to this extent it must also function as a *deflection* of reality" (45). Herein lies the theoretical description of a "terministic screen": by selecting the means through which we describe something, we have chosen a particular language format that necessarily filters out some suggestions while emphasizing others. I am reminded, for example, of a friend of mine trained in the field of urban development. When he sees an undeveloped plot of land, therefore, he looks at a tree in terms of how it can be removed from the land and used as a resource while concurrently making space to erect buildings. His terministic screen being informed by the dictates of a developer automatically forces him to describe the land in terms of how it can be used, thus emphasizing in his language how that can best be accomplished. A poet might see the same tract of land and describe it completely metaphorically, so that if I put the poet's description next to the developer's, I might not be able to tell that they were talking about the same thing. A Druid would probably discuss the land and the trees from a religious standpoint and would thus provide a third terministic screen through which the physical existence of the land might be filtered. The scientist might describe the land in terms of erosion, evolution, biology, and natural relationships (though he, too, would rely on the four master tropes to do so). Burke puts it this way: "much of what we take as observations about 'reality' may be but the spinning out of possibilities implicit in our particular choice of terms" (46).

This is a very useful idea if we are to understand the nature of Cold War scientific rhetoric. By viewing the Cold War as a particular set of terministic screens, ones that deflect certain realities while reflecting others, it is much easier to determine why one screen was favored while another was suppressed. It should be noted, however, that it is too convenient to view terministic screens as merely singular, like overlaying a transparency onto an overhead projector; rather, as in the case of the Cold War, there existed myriad transparencies at once, each emphasizing certain features while distorting others. Sometimes the screens even complement one another (much in the same way that ideologies do in an emergent culture), as in the case of the religion/science dichotomy, so that both the terministic screen of science and the terministic screen of religion underscore both features that are unique to themselves and features that they share. I am reminded of an anatomy encyclopedia I owned as a child, one section of which consisted of several pages of transparencies that represented the various anatomical systems: one for the circulatory system, one for the nervous system, one for the digestive tract, etc. When all the pages were laid together, one had the complete picture of the internal structure of the human body. Such is the nature of the Cold War: by laying the terministic screens on top of one another, we get a more complete picture of the social consciousness that drove attitudes during this period of American history. The difficulty, of course, is in separating the screens so that they can be learned individually after the whole has already determined many of the complicated perspectives that comprised the Cold War.

It should be noted that Burke's interest in science has been limited mainly to the so-called human sciences—those areas of study, such as psychology and sociology, which use a scientific mode of inquiry to discover facts about human behavior. Practitioners of the hard sciences—physics, chemistry, biology, etc.—often hold these fields in contempt because such scientists feel, not unjustly, that it is impossible to apply the scientific method to so inconsistent a subject as the human animal. That whole disciplines have been developed in an attempt to do this scientifically again speaks to our faith in science as a mode of knowledge; we feel that if it is possible to discern consistent patterns in natural objects with great precision, then we should have equal success if we apply these methods to human beings, their activities, and their institutions.

These ideas are important to keep in mind if we are to understand

some of the curious motives that drove domestic and foreign policy during the Cold War and how a receptive American public managed to
thrust itself into this perplexing way of thinking. If Michel Foucault is
accurate in his assessment of the appearance of science and scientific
discourse as an institutional entity, and I think he is, then there is much
to be said about how science influenced popular opinion regarding
nearly everything that touched public life in the 1950s. Ironically, faith
in science almost always overshadowed questions about it, despite the
fact that many of the problems encountered by the civic body could
trace the difficulties directly to a linguistic faith in science as an antidote
for all human struggles. It is and has been, moreover, a misguided and
unquestioning confidence that has perpetuated the myths about science,
and these myths have not often been discouraged.

A Way to Study the Rhetoric of Science

The theories covered here have much to say about the cultural preoccupation with science. Burke is interested in scientific language as a source
of motivational energy in our society, but other theorists, like Michel Foucault, are interested in science as a historical entity and a developing institution. Burke, always fascinated with the power of language to move
people to action, concentrates on how, specifically, this can be accomplished through the use of mere symbol systems. Others are a bit more
cynical in their estimation of language. Unable to pinpoint a central locus
of control over language, theorists such as Foucault concentrate on the
effect rather than the cause: the institutions that emerge and thrive as a
result of control over language and the distribution of terms that manipulate our lives.

A historical—or even archeological—study of scientific symbolicity
provides the tools to deal with language as a relic to be unearthed and
studied rhetorically. Given this, repressed and obscure texts provide
some of the most useful rhetorical evidence, because these are the ones
that have failed to make it conspicuously into our culture and our understanding of our own history. Language manifests and protects itself
in the great bulwarks of institutions such as science and religion, making it at once inscrutable and seemingly impenetrable to the people

whose lives it controls on the most intimate level. Such language is not impenetrable, but it *is* deliberately obfuscated for rhetorical purposes— purposes used by everyone, from the lowliest individual to the grandest social establishment. In short, by analyzing the institutions that are the result of a carefully executed orchestration of language, and by focusing on the language itself as the foundation for all symbol using and misusing as it is done on both the small and large scales, we can begin to uncover rhetorical patterns in the linguistic apparatus of science.

These ideas are important to our understanding of Cold War rhetoric. Such ideas deal with language as a basis for social formation. They articulate science itself as a language system and as a means for influencing cultural and intellectual values that determine human actions and reactions. They aid our understanding of science as a body of symbolic knowledge that affects the way we view our surroundings, and never from a strictly detached or neutral standpoint. They uncover the misappropriations of scientific power as it shapes the cultural and political framework of our society and, ultimately, of our civilization.

With such a theoretical substrate in place, it becomes necessary to use a critical method different from that conventionally used in a historical study. Whereas history itself is often burdened with the limitations of the scientific method, being primarily interested in basic statements of fact, geography, and linear events, I will combine the theoretical and critical strengths inherent in many disciplines: history, rhetoric, sociology, science, literature, and religion, to name a few. The aim is to demonstrate the merit of all these disciplines as critical tools and to show, ideally, how much richer a reading of our past can be as a result. If there is one unifying element to this interdisciplinary method, it is, as one might guess, rhetoric. The value of rhetoric as a critical apparatus stretches across the self-imposed boundaries of supposedly independent disciplines and helps reveal them all as constructions of language designed to shape our perceptions of past and present events. If the human animal does one thing universally, it is that it attempts to garner allegiance with its fellow human animals. This is always accomplished through language on some level, and rhetoric gives us one possible method for understanding the language we use. It is therefore supremely important that rhetoric lose the derisive standing it has inherited in recent times and be seen as the critical tool and comprehensive motivator it truly is.

Chapter 2

Precedents for Science as the Emerging Hegemony

As I have mentioned throughout the previous chapter, though science has become (in what we arrogantly refer to as developed countries, at least) the ideology that retains the most impact on our extensive public consciousness, it has not managed to sever itself from religion in any permanent way. The rhetorical features of both religion and science are too similar for them to ever become completely separate entities, and one reason for this is their ultimate goal of understanding ourselves and our role in the cosmos. Linguistically, both religious and scientific advocates draw upon the same master tropes, and often the same specific manifestations of these tropes, to explain themselves. But there is an important distinction: scientific discourse used to describe operations within its own sphere of study (a discourse reliant on mathematics, specialized language constructs, and restricted codes of communication) is designed to deliberately sidestep charged language formations. However, this is only one small element of science's rhetorical process. Prior to conducting the specified operations of experimentation, data collection, prototype design, etc., a scientist is just as prone to the influence of cultural value systems as any other member of society. A medical researcher interested in curing cancer will be morally (and therefore rhetorically) motivated by a social consciousness that dictates the need to ease suffering, pain, and disease in our fellow human beings. The entire Hippocratic oath is based in an ethical belief that a doctor's primary goal as a medical practitioner is to do no harm and to prevent harm whenever possible. An engineer interested in designing a cheap and re-usable spacecraft may be motivated by pure scientific inquiry and curiosity about the secrets space holds, but it is more likely that he or she is immediately concerned with something like finding untapped resources that can be mined for commercial gain. The contemporary truism that manned missions to Mars will be conducted by corporations, and not the government, reveals such motives in a transparent way. Even the governmentally endorsed space race of the 1960s, despite Kennedy's assertion that it was pursued in the interest of gaining scientific knowledge, was ideologically grounded in a political need to beat the Soviet Union to the moon. There are many such examples. The reason religion and science appear to do battle so emphatically has more to do with a cultivation of historical myth about the antithetical nature of these hegemonic bodies than we often assume. In order to understand why science and religion appear to be at odds, therefore, we need to examine some representations of their conflicting relationship in the past. This, in turn, will help explain the cultural heritage that informed the Cold War civic ideology—one that seemed to embrace a conflicting ethic of secular progress and modern manifest destiny.

As occurs in all intellectual movements, scientists had varied success in their attempts to bring their new and innovative ideas to acceptance, both with the powers that be and the laity alike. In this chapter, I draw a parallel between four key figures—two scientists and two philosophers—in history who had either success or failure in their abilities to persuade through the use of rhetoric. Galileo, whose talents as an astronomer are well documented, failed to understand the importance of carefully considered rhetoric and was, as a result, exiled from his native land and ordered to silence—never a very efficacious rhetorical strategy if one wishes to be heard. Francis Bacon, likewise, was so enamored with science that he proposed to overthrow everything that had previously passed for knowledge, and this highly ambitious enterprise led him to believe not only that science was capable of accomplishing this but also that anyone with a command of science could become godlike. The sheer extremeness of this position made Bacon no quick friends, and it soon became clear that he had misread both science's ability to accomplish this ambition and his audience's desire to achieve it. René Descartes, on the other hand, while a strong advocate of scientific and mathematical models, used this knowledge almost immediately to attempt one key proof: the existence of God. Because of his position that religion and science need not be battling for the same intellectual turf, he remained aloof from serious criticism and, one could even say, established the religious/scientific dichotomy in a way that has never completely left us. But the most able rhetorician of all had to be Charles Darwin, for though he privately discredited the creationist theory of human existence, he deliberately used the prejudices of this school to his own advantage, thus tapping into the ideology of the masses and the orientation of his detractors to reinforce his own theory. While his theory of evolution was rife with political controversy, he managed to maintain an even rhetorical stance, using the symbolicity of the authorities and god-terms of the laity in a way that allowed him to be heard on a topic no one was predisposed to listen to.

Even a passing examination of early science, especially during the Middle Ages, reveals its close ties with religion. Nearly all science during the Middle Ages was conducted by religious men: priests, monks, and novices of the various orders of Catholicism. Such an arrangement was inescapable since these people were the only ones who had intimate access to books, and the monasteries in which this segment of medieval society lived were virtually the only libraries in the West at the time. Only the members of orders such as the Benedictines or the Franciscans had the facilities, training, and leisure to read scientific texts and perform crude scientific experiments. This was a convenient arrangement and allowed members of these orders to establish doctrines that demonstrated a consistency between Christian thought and ancient knowledge. Furthermore, since Catholicism was more than simply a religion, but was also the central political authority, any discoveries were under the close scrutiny of the Church and required Church approval before they could be revealed to others. This could, of course, be a daunting task because of the Church's firmly established and closely guarded doctrines on the nature of the universe, which were generally an amalgamation of ancient thought and biblical Scripture. Anyone who challenged these doctrines could, at best, be severely ridiculed and, at worst, be tried and convicted of heresy, suffering the unpleasant consequences.

This was no small consideration for medieval scholars (which is perhaps a more accurate description than "scientists") who saw incongruities with Church teaching and the message of the ancients—one could be excommunicated from the Church for pressing an issue too hard, a fate worse than death for those who believed that the Church was the only path to salvation for one's eternal soul. Moreover, monastic monks did little in the way of actual science as we know it today; instead, they spent the majority of their time reading, transcribing, and sometimes translating existing texts from Greek to Latin, the official language of the Church. This was a tedious task but one ultimately indispensable to later scholars, since it provided new texts that were sometimes significantly divergent from the originals (if they still existed), thus providing clues to the mind-set of the Church authorities who oversaw the creation of new libraries and the books that were kept in them. The result of such a system was centuries of stasis—traditional interpretation and doctrinal stagnation that was incredibly difficult to overcome, if it ever even occurred to anyone to do so. The Middle Ages saw roughly one thousand years of looking at Aristotle in one particular manner (and this, classical scholars argue, using a corrupt Aristotle); it is little wonder that no one thought to question the authority of a tradition that was this old.

During the Renaissance we see the emergence of many great and influential scientists—Tycho Brahe, Giordano Bruno, Galileo, Leonardo da Vinci, etc.—who rigorously questioned existing dogma in order to find a more plausible truth in the operations of the universe. These were not the scribes of the medieval monasteries, but men who actually took on all the features of modern scientists by engaging in careful observation, formulation of hypotheses, and systematic testing of these hypotheses to discover whether they were in fact true. Earlier, Thomas Aquinas had fused Aristotle's philosophy into one with Christian dogma, so when the observation and experimentation of the fifteenth and sixteenth centuries began to cast doubt on Aristotle's physics, it was felt that Christianity was also being questioned since the geocentric theory of Aristotle had textual confirmation in the Scriptures (Dampier and Dampier 10).

One significant development in Renaissance inquiry rests on its increasingly secular nature: many of the scientists during this time were either members of the emerging middle class, who had the money and resources to conduct such experimentation, or they were commissioned by the ruling class, like the Medicis in Florence, who had financial interests in technological development. The result was that the men who practiced science were not under the immediate supervision of the Church but were instead privately funded and sanctioned, a situation that removed direct control from the hands of Church authorities. Copernicus (1473-1543) was one of the first to demonstrate the uneasy tension between Church and science when he revived the astronomic theory of a heliocentric universe in his *On the Revolution of the Celestial Orbs*. With the sun at the center of the universe, Copernicus was able to show that his theory explained the facts more simply than the cycles and epicycles of Hipparchus or Ptolemy. Though the theory was published with the consent of Pope Clement VII, the Vatican suspended it in 1616 because, it was proclaimed, the idea was "false and altogether opposed to Holy scripture" (10). This event would be only one among many in a series of conflicts between the Church and the scientists of the Renaissance.

Another, perhaps more famous example of the tension between science and religion can be seen in what is sometimes referred to by historians as the Galileo Affair, an event that demonstrates two things: the increasingly strained relationship between religion and science in their pursuit of control over intellectual primacy, and the distending role of rhetoric to determine the outcome of this intellectual battle. It is interesting to note, however, that the Church was perhaps not as ignorant and unreasonable as popular historical myth would have us believe. In the mythological version that has been passed down to us, science and religion are seen as ideological nemeses, and only through painful deliberation was science able to liberate itself from the intellectual prison of medieval theology and become the shining harbinger of truth that Galileo was able to make it. However, according to George Sim Johnston, it was Galileo who had become so intractable in his position that the Church was forced to exile him in the end in order to preserve Church authority framed by a geocentric model of the universe (1).

This story requires some further historical background. The prevailing attitude regarding science during Galileo's time was that the purpose of astronomy was to save the appearances of celestial phenomena—that is, to take measures to preserve the Ptolemaic system because it was compatible with Scripture. However, the Ptolemaic system was, according to Johnston, simply a mathematical tool that medieval astronomers had at their disposal—a tool which gave a numerical order to the cosmos. In other words, there was no overt attempt to save the Ptolemaic system on religious grounds and at the expense of scientific accuracy; it was simply used because it had been used successfully for centuries on grounds that were ultimately removed from a quest for absolute cosmic truth (2).

The Ptolemaic system was, however, intellectually burdensome, so when Copernicus reintroduced the heliocentric theory (Aristarchus of Samos was probably the first to make this proposal), a much simpler tool to describe the motions of the heavens, scientists like Galileo were more than happy to adopt it. When Galileo began using the telescope (which was invented anonymously in Holland, according to Johnston, and not by Galileo as history frequently teaches), he made some startling discoveries: the moon was not a perfect sphere, an observation which contradicted the doctrine of perfect celestial bodies; Jupiter had at least four satellites, which contradicted the notion that heavenly bodies revolved exclusively around the earth; and Venus had phases, which meant that it must revolve around the sun and not the earth. Once these discoveries were made, Galileo showed his findings to the leading astronomer of the day, a Jesuit named Christopher Clavius, and it could not be denied that Galileo was in fact correct. The result for Galileo was that in 1611 he was transported to Rome and given a private audience by Pope Paul V, who assured him of his support (3).

This event more or less ended Galileo's desire to conduct pure science. According to Johnston, he seemed much more interested in converting public opinion to the Copernican system, which suggests his propensity toward rhetorical persuasion. Unfortunately, Galileo was a much better scientist than he was a rhetorician—he managed to systematically alienate anyone who would not wholeheartedly accept his idea. It apparently never occurred to him, or seemed simply irrelevant, that it might be difficult to overturn centuries of tradition that was well supported in the Scriptures of Christianity. As Johnston puts it, Galileo "was intent on ramming Copernicus down the throat of Christianity" (4). Oddly enough, the Catholic hierarchy had given him much more than the benefit of the doubt and was willing to consider his ideas in good faith, but he was so acrimonious in his position that he left the Church authorities with a virtual ultimatum: they must either accept Copernicanism as a physical fact and reinterpret Scripture accordingly, or they had to condemn it as heretical. The Church even offered to consider Copernicanism as a hypothesis, perhaps superior to the Ptolemaic system, until further proof could be provided (5). Galileo would have none of it.

Through a series of events (the details of which I will not list here), Galileo was brought before the consultor of holy office, a.k.a. the Inquisition, no less than four times and was put on trial twice. In the last trial, Galileo was sentenced to "abjure the theory and to keep silent on the subject for the rest of his life, which he was permitted to spend in a pleasant country house near Florence" (6). This was unprecedented leniency, especially considering that Galileo had been so inflexible in his approach and abrasive in his rhetorical tactics over the last couple of decades of his life. Though he was officially condemned as "vehemently suspected of heresy" (from which the legend of his crusade undoubtedly stems), this charge was more a way of making an example of Galileo than any absolute condemnation of his theories. In short, the Church was willing to entertain Galileo's Copernican system, but he was so insistent on having the entire Western hegemony stood immediately on its ear that the Church was left no other choice but to have him gently, if effectively, silenced.

In the case of Galileo, we can see how the myth of religion and science being at odds became part of our cultural orientation; it seems almost necessary for us to convince ourselves (scientists and laypersons alike) that science is a clear and progressive divergence from religion. After all, where science is based in reason, fact, and rational intellect, religion is often engrossed in superstition, ceremony, ritual, and faith. Where one champions an unfettered pursuit of the truth, the other relies on its followers to make incredible leaps of faith that, on the scale of reason, seem to justify the otherwise unjustifiable. Where one relies on painstakingly gathering data to produce results, the other relies on ancient texts, belief, and tradition to form its substance. Clearly there are distinctions between the two, but the differences are in the approach rather than the goal; both science and religion wish to secure and understand the dynamics of human activity in and relationship to the universe. The story of Galileo demonstrates that the differences between science and religion were insurmountable at that time, and this is how we have now come to view these seemingly disparate bodies of knowledge. Yet we have come to a point in human history where science has not only acted as a substitute for religion in many cases, but is actually fulfilling the ordering role that religion once occupied. This role can be described, if you will, as the mouthpiece of human fate.

It is important, for example, to note that the first modern scientists were given credit for, at least so far as we have lent them historical credence, their efforts in astronomy. If Copernicus and Galileo can be considered the first modern scientists, why is it that they had such a preoccupation with a discipline that was not their first calling? Galileo's first experiments were in physics, not astronomy, and while these two disciplines are clearly interlinked now, they were certainly considered less compatible then. One possible explanation is that these men desired a status equal to or even surpassing that of the ecclesiastical body—to provide explanations for the creations of God in a way that had eluded centuries of Catholic doctrine. The interest, if this is true, was not in "science for science's sake," as we like to believe today, so much as it was in a desire for universal knowledge and the privileges that would accompany this status. The fact that Galileo effectively abandoned his scientific pursuits after his discoveries in favor of his rhetorical campaigning, if Johnston is correct, suggests that he wanted the highest recognition of the papacy more than he wanted to be a scientist.

Philosophically, I should note that one of the underlying objections to Galileo's discovery was not only that it contradicted the divine ancients and Scripture but also that it was a heavy blow to human egocentrism: mankind was being neatly removed from the center of the universe through the heliocentric system. While some have argued that Catholic doctrine insists that man is not the cosmic center, but rather one element in the base physicality of the divine system, the scriptural arrangement of heavenly bodies belie this claim. Mankind may have been viewed as base and flawed, but he was still the divine purpose behind the creation of the universe, so far as the Church was concerned. A heliocentric process that placed man on the outskirts of a much larger and considerably more indifferent natural order might be described, metaphorically at least, as a blow to religion from which it might never fully recover, and Galileo wished to gain credit as the new guardian of rational knowledge. The Catholic Church had already lost considerable power and influence through its recent bouts during the Reformation; it was probably not ready to willingly concede further blows to its authority by automatically granting Galileo his pet project.

The myths surrounding the events of Galileo and his conflicts with the Church seem to support the popular notion that science and religion are at opposite ends of the ideological spectrum that represent two disparate modes of inquiry. I bring this up mainly to show that, almost from the beginning, science and religion have had an at once symbiotic and conflictive relationship, an arrangement that would evolve over time just as each has evolved individually. In short, science necessarily sprang from the head of religion, and we still treat science as a form of religion when we turn to it to answer the most pressing of our cosmological questions. It is imperative to understand that, despite the common misconception that religion and science are natural ideological adversaries, they in fact not only share common features with one another but are also born from the same human need to know—the constant struggle for finding certainty, consistency, and faith in the world around us. Our conviction in science is no less a product of our desire than our faith in religion, and both are, as I will continually remind the reader, value-laden means of imposing order onto chaos, sometimes to our own detriment.

Philosophical Implications: Bacon and Descartes

Thus far I have been speaking of science as a body of knowledge that is ideologically pregnant with the human desire to discover Truth. However, science is, and has been since its earliest inception, chiefly an orientation concerned not only with a pursuit of Truth but with the concomitant desire for power: power over the natural world, materials, one's own behavior, and other individuals. It is no accident, for example, that nearly all technological discoveries have had their first application in weaponry or the military, as if the human animal is motivated at its core by a need to dominate and control its surroundings and other people. Science, as a strictly pragmatic methodology, what Bertrand Russell calls the "scientific outlook" (see *Science and Society*), creates a terministic screen prone to exploitative applications since this is the nature of its process—to study, manipulate, and control the natural world.

But this, of course, is not a satisfactory definition of science. Is science merely a vehicle through which human beings can dominate, or is there more to this powerful system of thought and practice? Is science, after all, a *philosophy* of thought, with all the moral, intellectual, and metaphysical connotations that philosophy implies? I have looked at some of the early developments in science as a practice, but we must also look at science as a principle of thought. How did science take hold, and in what way did it exhibit itself in the early modern mind? Philosopher W. T. Jones offers a possible insight to this question:

> Insofar as it was a matter of questioning old authorities, the development of scientific method was simply a part of the widespread movement of thought reflected in the transformation of political and moral ideals and the weakening of the Church's hold on society. Insofar as it was specifically an interest in nature, the empirical spirit manifest itself, for instance, in Petrarch's appreciation of the view from Mont Ventoux, and in Chaucer's realistic portraits of his fellow pilgrims and his delight in the sounds, smells, and sights of spring. It was also at work in the growing passion for exploration that sent men in search of new routes to the Indies, as well as the concern for correct rendering of the human figure and the devotion to problems of perspective shown by the painters of the fifteenth century. (68)

We see that, early in its evolution, science was becoming a manner of thinking, a world view, and a dominating terministic screen that carried with it more than the sum of its methodological parts; it was, in fact, becoming a dominant *Weltanschauung,* one that colored the way in which people would see the world as a whole and their place within it.

Francis Bacon (1561-1626) is one of the earliest examples of a thinker who gave himself over to the notion of science as a means to power; he is one of the first to ingest science as not only a practice but also as an ideology. While he preferred to think of himself as having ambitions only in the intellectual realm, he was far from the objective experimenter interested only in gathering and ascertaining facts and data. As Jones puts it, "It was not that he was uninterested in 'the Truth,' as he called it; what interested him about [science] was chiefly its possibilities for exploitation. What he sought, in his own pregnant phrase, was 'that knowledge whose dignity is maintained by works of utility and power'" (75). While history provides examples of men whose primary motive was power—Cesare Borgia, Louis XI, and Machiavelli, for example—Bacon distinguished himself by realizing that science was the most efficacious means to this end. Bacon would accomplish his fortune, as he puts it, through a "total reconstruction of the sciences, arts, and all human knowledge" (75). He called this scheme the "great insaturation," and it held two basic premises: (1) everything that had up to that point passed for knowledge was in error; and (2) the human mind was an inadequate instrument for obtaining knowledge (75). The human mind was, then, in a degenerative state that differed from its original, pristine, and native condition.

The interesting implication of this idea, rhetorically speaking, is that where the prelapsarian metaphor was once used to describe the moral condition before the Christian Fall of Man, it was now being used as a secular description for the state of the human mind. The idea, as Jones puts it, created a kind of "epistemological Protestantism" that held "just as man's original, moral will had been corrupted and had become enslaved to sin, so his original power of knowledge had become diseased and had led him into sins against the 'Truth,' namely, the acceptance of all sorts of errors and mistakes" (76). It was more important for Bacon to ask, as did Luther, not how the Fall occurred, but how we could be restored to our original, pristine state of mind, a mind that would be empty of all preconceived notions and cultural snares. While the philosophical problems with Bacon's idea are clear (a mind empty of all preconceptions would be incapable of discerning the experiences necessary to analyze the facts), the coupling of theology and science is what should interest us here. Once again we see that the principles of religion are impossible to shake off, even when describing so seemingly discordant (when compared to Christianity) a system of thought as science.

As for Bacon's philosophy, his main criticism of medieval science was that it was deductive and therefore not purely a natural science. That is, the science practiced by medieval scholars consisted primarily of words used to describe generalities rather than particulars; Scholastic explanation (which was often considered a science by the scholars of Bacon's day and had been throughout the Middle Ages) was based on an Aristotelian notion of potentiality in form and matter using syllogistic logic, a system that, by Bacon's estimation, failed to account for the facts by using empty verbiage instead of rigorous examination: "The syllogism consists of propositions, propositions consist of words, words are symbols of notions. Therefore if the notions themselves (which is the root of the matter) are confused and over-hastily abstracted from the facts, there can be no firmness in the superstructure" (Jones 77). In short, Bacon distrusted a system that was too reliant on the arrangement of words and not on the how the major premise of an argument corresponded to the facts in question. In order to counter this problem, we must expose false notions and prejudices that are prone to hasty generalizations and prevent us from seeing their exceptions; that is, we must rely more completely on empirical data and divorce our minds from the language biases that might prematurely shape our analysis (78).

Another important consideration here is the apparent divorce of language from scientific analysis, a separation that lays a precedent for later scientific thought in that it assumes two things: that scientists are not fettered by the illusion of symbolic language, and that, because of this, they can be entirely objective when applying an inductive method when doing science. Bacon was convinced that a good scientist could clear his mind of the linguistic baggage that had plagued the Scholastics, but this was a naive assumption, since in order to practice science, one must have a language with which to both analyze and explain the findings.

The language of preference among scientists, which stems from this concern, therefore, would be mathematics. And this brings us to René Descartes (1596-1650). In Descartes's autobiographical work *Discourse on Method,* he describes an intellectual bildungsroman that captures an acute need for certainty in his philosophical pursuits. This desire for order and certainty would be a defining feature not only of his own particular philosophy but also of science in general. One interesting detail of Descartes's philosophical pursuits is his description of a mystical experience that, instead of leading to the salvation of his soul, led to the discovery of a new scientific method (Jones 155). Here, yet again, we see the similarities between religion and science, but also the differences. Where Descartes, like Augustine before him, had a vision that changed the course of his life and, one might say, his faith, the direction of that course pointed to the human intellect instead of God. Descartes, unlike Bacon, had the utmost confidence in the human mind to solve all earthly problems, and this too would become a defining characteristic of science, especially during the Cold War. In the sense that the human mind was capable of extremely complex operations, it seemed likely that human beings were destined to achieve the status of gods. One might say that this faith in the human intellect motivated scientific pursuits in a way that is still documentable today.

More importantly, the mystical vision that Descartes claims to have experienced was, he believed, given to him by God so that he could understand the "foundations of a wonderful new science" (156). He was, he felt, given a privileged insight into the workings of the universe (and the mind of God) that could only be fathomed by using the gift of science. The method of scientific inquiry itself was a compilation of twenty-one *Rules for the Direction of the Mind,* which, again, reflect a passionate desire for certainty and a stalwart faith in the human capacity to achieve it. Even theology could be defined as knowledge of revealed truth, a distinction on which Jones comments that in "an age of secure faith, insistence on a sharp distinction between philosophy and science had seemed to insure the tenets of religion. But now, in an age of reason, abandonment of the Thomistic notion of a natural theology merely debarred theology from an effective share in the practical affairs of life" (158). Descartes believed in the replacement of medieval Scholasticism with a "secular instrument" that would be coined "reason." Reason was a universal instrument because it could perform, so it seemed, any task placed before it (159).

The nature of reason, though held as the fabric of the empirical age of Descartes, was not universally agreed upon. For Descartes, reason required that the mind be turned inward upon itself so that it could seize upon some self-evident truth. The primary basis upon which Descartes operated was mathematics, since he had always been fascinated by the certainty that it was able to demonstrate. The thing that made mathematics certain for him was its ability to reflect what he believed was the rational order of the world. If math was the most "clear and distinct" language at the disposal of the human mind, then it made sense to use it as a way to erect a metaphysical structure complete with an ethics and a means to prove the existence of God (160).

Significantly for the purposes of science, mathematics (in most cases) is a strictly linear symbolic system, which makes numbers sequential by nature and, therefore, predictable in behavior. The impact of linear, numerical thinking in science is a topic that would be an interesting study in its own right, but suffice it to say that because of the mathematical models that science has adopted, science is required to think on a causal, inductive level—a level that has many of its roots in Cartesian metaphysics. The use of numbers as a basis of thought is, at the risk of resounding the pejorative, minimalist and reductive. While we may quibble over the definition of liberty, justice, or freedom, the properties of 3, 81, or 1,987,625 are far less ambiguous and, in much the same way as Scholastic logic, harder to dispute. This degree of certainty as a defining characteristic of mathematical operations will embed itself in modern thinking, and our quest for it will help buttress such inflexible ideologies that aided in the construction of historical events like the Cold War. As a language system, math has all the advantages of neutrality, since while one may associate many positive and/or negative feelings with a word like *love* (especially depending on its context and its relationship to other words—see I. A. Richards's *The Philosophy of Rhetoric* for an explanation of the interinanimation of words in their various symbolic and textual contexts), it is far less likely that one would have equally intense reaction to, say, the square root of 127.

Descartes felt that in order to provide certainty in metaphysics, one must search for the same absolute principle that one might find in, say, geometry—that is, a theorem indisputable and self-evident. This primary principle, if it could be discovered, would yield an objective and rational—scientific—reality. In order to do this, Descartes produced the now-famous method of systematic doubt that held that everything—every belief, value, or standard—must be challenged, no matter how accepted or plausible it may be. The method was simple: suspend beliefs until they could be proven conclusively. He accomplished this by finding an absolutely certain starting point and then reconstructing his beliefs as they were deduced through the geometric method he had developed (162). This method is Cartesian skepticism, and it led to the only principle through which Descartes felt he could have absolute certainty: *cogito, ergo sum* (I think, therefore I am). It was, ironically, the doubting itself that had led him to this conclusion. He knew that the only thing certain that could be gleaned through his new method was that he doubted, and doubting was, of course, a form of thinking (164).

Because Descartes was a devoutly religious man (which goes far to explain why he had attempted to create a metaphysical system distinct from medieval Scholasticism), he needed to account for the existence of God. His proof for the existence of God can be broken down logically in this way: (1) everything, including our ideas, has a cause; (2) we have an idea of God; (3) nothing less than God is adequate to be the cause of our idea of God; (4) therefore, God exists (165). This proof is, actually, a modification of Anselm's ontological argument in that both Descartes and Anselm held that the idea of God necessarily has God as its object. The problem with the proof, however, is twofold. First, we must ask whether we really have a clear and distinct idea of an infinite and perfect being, and, second, we must ask whether God is its only possible cause (167).

Aside from the theological complications with Descartes's proof, it is important to ask why a man so preoccupied with reason, order, and certainty and so dedicated to the reformation of Scholastic knowledge would trouble himself immediately with the pursuit of spiritual questions. The answer is simple: while the methods employed in the past were inadequate to answer such questions, the questions themselves had changed little. And the question preoccupying humankind was always where, how, and why we belonged in the universe. While Descartes had gone far in recasting methodological inquiry, he could not help asking the same questions that had been asked for centuries. Since he *did* crave certainty, it is only natural that the first thing he would want to be certain about was the existence of a supreme being.

Descartes's desire to prove the existence of God does not detract from his enthusiasm for the new science; as a loyal Catholic, he was well aware of the Church's misgivings about the implications of the new science and he was therefore motivated by a desire to show that such concerns were ill-founded. This way, he could free scientists from Church interference that was, Descartes knew, slowing down scientific progress. As a result, Descartes developed a compromise that held "if mind and body are completely different kinds of things, and if the truths about each follow from the distinct nature of each, it is possible for the science of minds and the science of bodies to contradict each other" (Jones 176). The implication, therefore, was that theologians have no reason to interfere with physics just as physicists cannot claim any special competence on spiritual matters. Science and religion are not in conflict because each is dominant in its own sphere and neither has any standing in the sphere of the other (176). More importantly, the Cartesian Compromise set a precedent of noninterference that would help define the religion/science relationship for following generations.

The significance of Descartes for this study is one of emphasis: where previous philosophers, scientists, and theologians had battled for their own intellectual turf, Descartes was one of the first thinkers to recognize the importance of both the new scientific method and the need for spiritual reinforcement. Descartes established a mind-set in the thinking of both religion and science that is still with us today. His notion of "clear and distinct ideas" is still a guiding principle in scientific inquiry, and his "compromise" created some intellectual boundaries between science and religion that are rarely deliberately breached. However, if we see in Descartes the source for our attitudes about the difference between religion and science, we should also see him as a source for their similarities. Before Descartes, it was assumed that if science made a discovery that contradicted Scripture there was automatically a conflict of interest with the Church. Descartes attempted to show that these two bodies of knowledge were sufficiently distinguished that they need not be at odds with one another. While Descartes went far to ease the tensions between the Church and science on an intellectual level, we should remember that the connection was never completely severed. If anything, Descartes managed to create an illusion that the motives behind religious pursuits and scientific pursuits were different, when in fact they are essentially the same. Where his influence is seen in our modern assumption that science and religion never need meet intellectually, what he really did was mask the simple truth that both seek the same end, namely, to discover a definitive order to the world.

Darwin and the Roots of the Evolution/Creation Controversy

During the 1700s we see the emergence of what is today referred to as natural history, a branch of biology that prefers to interpret present natural conditions through the evolutionary evidence that is written, for lack of a better metaphor, in the text of the earth and the creatures that inhabit it. This school of biology did not begin, as often supposed, with Charles Darwin and his pioneering work, *The Origin of Species,* but had been practiced for at least one hundred years prior to Darwin's journey on the *Beagle* (1831-36) to the Galapagos Islands. Naturalism, as a school of biological science, is important because it, like the cosmological observations of

Copernicus and Galileo, threatened the existing Christian assumptions by contradicting Scripture and age-old traditions in the understanding of not only the origin of life on earth, but, more important, the origin and significance of human beings. Where the Copernican heliocentric system had removed humanity from the center of the universe, naturalism threatened human significance even further by suggesting a rather arbitrary (and certainly not *necessary*) natural emergence of Homo sapiens as a species, putting us on par with all the other lowly beasts that occupied this planet.

To speak in Kuhnian parlance, the school of Naturalism had acquired two paradigms by the mid-eighteenth century. One of the naturalist paradigms was Linnaean (after naturalist Carl Linnaeus) and the other Buffonian (after the French biologist Count de Buffon), and they were, according to John C. Greene, "diametrically opposed in spirit, presuppositions, and concept of scientific method" (8). The former was a blend of Aristotelian logic and teleology but embraced a static form of Christian creation that identified natural history with taxonomy. The latter used a Cartesian notion that suggested nature was a self-contained system of matter in motion, and that by observing uniformities in the effects of this hidden system, it could construct models capable of explaining the observed effects as necessary outcomes of the system (8).

Darwin already had established paradigms on which to draw for his own theories, paradigms that were at odds with one another for both scientific and religious reasons. Where Linnaeus's theory can be seen as merely descriptive (in a way that is less threatening to existing religious doctrine), Buffon's theory (the one which Darwin draws upon most heavily) provides a more dynamic, causal explanation for the broad variety of life. For example, Buffon felt that the uniformities observable in nature were not evidence of a system designed by a wise and omnipotent Creator; instead, they suggested an extension of Newtonian matter in motion—the same theoretical system that provided an explanation for the evolution of the solar system (8). Such a system was clearly reliant upon the contingencies of an evolving planet and could not be said to be the product of any preordained, omniscient plan. In the Linnaean system, what Greene calls the "static paradigm of natural history," species had been defined as part of the stable framework of creation, which maintained that all species always had been as they currently were. One obvious problem with this theory was that it made no allowances for eventualities such as extinction or adaptation. If this were allowed, it could mean the unraveling of the entire system, and thus, of life altogether. As a result, many naturalists of the Linnaean school were reluctant to concede that a species could in fact cease to exist or adapt to their surroundings over time (10).

That either the Linnaean or the Buffonian paradigm might have gained favor had it not been for Darwin suggests that the connection between science and religion, even in the nineteenth century, is a mutually dependent one. And Darwin himself, through the struggle to gain acceptance of his theory, as we know, was certainly not without his religious detractors, many of whom were themselves scientists trained in the Linnaean system. The basis of Darwin's theory, of course, was the concept of natural selection, which held that if one species was best suited to its surroundings because of its own natural protection, adaptability, etc., that species would survive to reproduce another generation which preserved those favorable variations and attributes. This theory not only worried creationists of the strictly religious sort but also caused a general upheaval in popular thought, for even the most casually faithful realized the implications of creationist interpretations for the existence of earthly life.

Some discussion of Darwin's work may shed light on the radical nature of his theories and help explain why he was received with such suspicion early on in his attempt to prove his ideas. In his 1871 work *The Descent of Man*, for example, Darwin describes in candid detail the biological condition of human beings. The title of the opening chapter, "The Evidence of the Descent of Man from Some Lower Form," indicates at once how his theories were antithetical to the creationist paradigm by suggesting that Homo sapiens has not always existed in its present state but has instead evolved from creatures less physically and mentally developed. He cites as evidence the similarity in bodily structure, reproductive process, embryonic development, rudimentary organs and hair, and brain development between humans and other advanced mammals. From this he concludes the following:

> Thus we can understand how it has come to pass that man and all other vertebrate animals have been constructed on the same general model, why

they pass through the same early stages of development, and why they retain certain rudiments in common. Consequently we ought frankly to admit their community of descent; to take any other view, is to admit that our own structure, and that of all the animals around us, is a mere snare laid to entrap our judgment. . . . It is only our natural prejudice, and that arrogance which made our forefathers declare they were descended from demi-gods, which leads us to demur to this conclusion. But the time will before long come, when it will be thought wonderful, that naturalists, who were well-acquainted with the comparative structure and development of man, and other mammals, should have believed that each was the work of a separate act of creation. (Darwin 246)

The reference to the structure of humans and animals as "a mere snare laid to entrap our judgment" is an especially apt rebuttal to the religiously based arguments against his theories, which claimed just that: that the world as it was viewed by man, and the discrepancies contained within it on geological, biological, and environmental levels, were simply tests of faith from an omniscient creator who knew that we might someday question our origin. Such an argument, by today's standards, might seem ludicrous; after all, what kind of god would intentionally try to fool us into *not* worshiping him? Nevertheless, this was precisely the counter provided by the creationist faction in an attempt to discredit Darwin's theories. That Darwin recognized this and spoke rationally against it (as in the last sentence where he calls each of us "a separate act of creation") suggests that he is, in fact, a clever rhetorician. While he will not capitulate to the "divine trap" argument, he will draw on the language of his detractors to make his own. By associating his own theory with that of his opponents through the use of the word *creation*, he has urged them to consider his argument both on its own merit and in relation to theirs. It is not one grand creation that produced humanity, but something far more impressive: separate, individual, and myriad acts of creation that happen every day. Rhetorician John Angus Campbell, in fact, argues that Darwin, as much as any scientist, was as adept at rhetorical persuasion as he was at biology. Attention to his rhetoric does not detract from his abilities as a scientist, but draws "attention to his accommodation of his message to the professional and lay audiences whose support was necessary for its acceptance" (69). The fact that *The Origin of Species* was made available to the general public as an abstract is one indication of its rhetorical nature. Its brevity is another. The ethos of the author—that Darwin appeals to the reader's sympathy through passages such as "my health is far from strong" and "I must trust the reader reposing some confidence in my accuracy"—provides further evidence of its rhetorical intent (70). Even the use of themes such as "origin," "selection," "preservation," "race," "struggle," and "life," Campbell claims, shows a colloquialism, an intimate design, that is at its base rhetorical (70).

Campbell also points out, quite significantly, that while the theological objections to Darwin's work are well documented, it is not as well known that *The Origin* also contains a theological defense within it, particularly in its treatment of natural theology. According to Campbell, who cites several examples, the body of *The Origin* pays proper tribute to the ways of Providence over the views of his opponents (71). Still, the evenness of Darwin's tone, and his apparent respect for the beliefs of the Christian faith, suggest that his audience, both popular and professional, was in the author's mind during the crafting of the book. By appealing to the audience's deepest beliefs—an appeal that appears genuine—Darwin has established a much more sympathetic reaction to views that might otherwise fall on deaf ears because of those deep beliefs.

Conversely, Darwin also writes the text from a standpoint of common sense, urging us to trust a theory that explains many sweeping facts about the common events of everyday life. The appeal to common sense is always a compelling rhetorical strategy, especially in an age that strongly embraced the virtues of reason, rationality, and logic. By balancing the tenets of faith with the rules of common sense, Darwin has managed to tap the two most potent value systems of the nineteenth century (religion and scientific inquiry) and again has accomplished an ethos difficult to dismiss.

Yet another interesting rhetorical feature of *The Origin* is its literary style. While its is important to note that eloquent prose was more of a standard in the scientific texts of the nineteenth century, if the artistry became too conspicuous, readers (especially professional scientists) might become suspicious of the substantiality of the argument. Darwin managed to avoid this problem by explicitly de-emphasizing both the colorful nature of his prose and the theoretical impact it was designed to express. This is a method Campbell refers to as the Mark Anthony Effect, a trope through which the speaker denies abilities at eloquence while proceeding to deliver an eloquent speech. Such a device helps disarm the imposing nature of conspicuous eloquence, leaving the audience more at ease to receive the content in a persuasive manner. In this way, Campbell notes, "rhetoric is freely employed and effectively masked" (73). Also, despite his caution with appearing overly literary in his delivery, Darwin drew consistently on metaphors to express his ideas. In this way he was able to describe complex, quite alien biological theories in terms that could be consumed by a popular audience. He managed this by, again, de-emphasizing the import of the images he used and, consequently, was able to direct the reader's attention to the substance of the metaphor without drawing overt attention to it (77).

This method, however, was not as successful with the scientific community. Darwin was frequently criticized by his peers for relying too heavily on imagistic language. To this, Darwin had a very legitimate rebuttal: "In the literal sense of the word, no doubt, natural selection is a misnomer; but who ever objected to chemists speaking of the elective affinities of the various elements?—and yet an acid cannot strictly be said to elect the base with which it will in preference combine. . . . Everyone knows what is meant and is implied by such metaphorical expressions; and they are almost necessary for brevity" (qtd. in Campbell 78). Such a view of the connection between the linguistic and the scientific is rare. Darwin clearly realizes, as evidenced by this passage, that regardless of the body of facts we have at our disposal, how they are received is as much a matter of expression as it is of content. Without the kinds of metaphorical associations used by Darwin (the very term *natural selection* was itself suspect because it implied an autonomous will on the part of nature), we could have no clear understanding of the theory nor any interpretive tool to use with the data collected. While the use of a metaphor, as opposed to literal language, must necessarily have some shortcomings in precision, it is a necessary component for introducing a theory as radical as Darwin's; without metaphor, there is no known condition with which to compare the phenomenon Darwin describes.

Darwin's theory itself is a bit more difficult to assess rhetorically, though it is clear that he made such an attempt. The difficulty, again, was with presenting his theoretical model in a manner that did not diverge too severely from the methodological conventions of science at the time. For example, it was generally held by the scientific community that the deductive model of observation was more useful than its counterpart induction, since one must have a theoretical framework to base any observations on; if one had no working theory on which to apply observations, the observations were merely facts recounted in a theoretical vacuum. Darwin, through his personal journals, showed the value of the inductive method for arriving at innovative new conclusions. Publicly, however, he scoffed at induction as a rather uptight, self-defeating scientific process because it seemed to merely practice observation for its own sake. Some have called Darwin a hypocrite for this discrepancy. Campbell makes the valid claim, however, that Darwin understood the value of publicly endorsing the status quo in order to gain a fair audience, something he did for strictly persuasive purposes in *The Origin*. This, along with many of his other rhetorical strategies, was done in order to "minimize the shock of novelty *The Origin* would occasion" (75). It is important to note, as Campbell does, that "no one serious about making a revolution can lightly ignore accepted professional standards. How far one goes in deferring to standards irrelevant or hostile to one's actual research procedures determines the personal dimension in science" (76). In short, early drafts of Darwin's work did not conform to the deductive model of scientific methodology, but he was careful to represent his views as if they did (76).

This brings Campbell to the conclusion that Darwin's philosophy of language was as important to his scientific innovations as it was to his popular and professional success because this philosophy, "rather than the . . . positivist theory of language to which he publicly deferred, helps explain his success in establishing a novel research paradigm" (79). One might say that Darwin's approach to language mirrored that of his approach to conducting science: in both cases, he was concerned with the advancement of an idea over the conventions that kept such ideas from thriving. More important, he seems to be one of the earliest examples of a scientist enlisting public support to help further his project. He was, in the words of Campbell, a rhetorician as well as a scientist who helped act as the "bridge uniting science with culture" (84). This bridge would become very important, for in the decades to follow, the relationship between scientific advancement and public interests would become increasingly close; so close, in fact, that they could not be said to exist in quite the same way without one another.

The purpose of this chapter was not only to provide some historical precedents for science as a dominant ideology in the West, but also to show that the nature of the scientific hegemony is at its base language oriented. The discussion of the emergence of science, with its paradigm shifts and religious similarities, was designed to show that science is at its core capable of eliciting the same faith that people once lent to Christianity. Further, as scientific endeavors continued to gain favor and popularity, science became increasingly tied to rhetoric in order to sustain this status when introducing new ideas or controversial theories.

Since the word ideology is used in a variety of contexts and from a variety of sources, it seems prudent to reiterate how I am using it here. While Marx had as specialized a grasp on the term as anyone, I am more inclined to use Kenneth Burke's modification of Marx's phrase. In *Counter-Statement,* Burke refers to ideology as a "vocabulary of belief" (146), a substratum of language that invokes emotion, attitudes, judgments, even ethics. Burke tends to use this term in conjunction with the literary notion of form—that is, as a means through which a writer can construct a text to appeal to the ideology as he or she sees fit. Ideology is, in essence, language that is rhetorically exploitable in an audience to enlist support by conjuring associations, either positive or negative. In *Language as Symbolic Action,* Burke provides a more analogical explanation: "An 'ideology' is like a god coming down to earth, where it will inhabit a place pervaded by its presence. An 'ideology' is like a spirit taking up its abode in a body: it makes that body hop around in certain ways; and that same body would have hopped around in different ways had a different ideology happened to inhabit it" (5).

In order for this to make sense for the present study, however, we must extend the definition of *text* to include not only the written documents of fiction, nonfiction, essays, or poetry, but also visual media, popular vehicles such as movies, television, and computers, classrooms, sermons, political speeches, newspapers, and magazines. Any or all of these mechanisms can be considered texts that further particular ideologies, especially those that lean more toward a visual, nonliterate format, since it is becoming much more commonplace for people to receive information, training, and ideas from these sources.

Which brings us to the test case for my theoretical claim: that scientific rhetoric was used to garner public alliance in the development of nuclear power (among other scientific projects) during the Cold War. The next chapter discusses how the images of nuclear power and weaponry, established very early in the twentieth century, took on various and sundry meanings as the nature of the conflict, and the expectations of a nuclear war, became increasingly real in the minds of Americans in large part because of changing ideological assumptions, but, more important, because the evolution of American science and technology supplied a means to initiate and buttress such assumptions.

Chapter 3

The New Age of Science

In the September 1950 issue of *Scientific American*, J. Robert Oppenheimer wrote the following: "We hope ... for agreement and understanding from an increasing number of men who are not scientists, but who are nevertheless concerned that advances in science make the greatest possible contribution to human welfare" (21). This statement appeared in an article by Oppenheimer surveying scientific accomplishments that had spanned the past fifty years. The year 1950 was more than a symbolic landmark of the mid-twentieth century because it had already seen some of the most massive scientific advances in human history. Oppenheimer, quick to showcase the progress enjoyed through the fruits of scientific knowledge, was equally mindful of the "errors and byways" of such progress. In his introduction to a series of ten reports written by *Scientific American* on the state of science in America, we might note that his tone is one of careful enthusiasm tempered with necessary caution. As a man instrumental in the development of the atomic bomb, Oppenheimer was in a unique position to see the double-edged sword that was scientific knowledge—Oppenheimer was keenly aware of the fact that two world wars had played no small part in this advance: "Yet at this hour in history one cannot read these 10 reports, which constitute so substantial an account of heroic human achievement and so persuasive an example of the progress of civilization, without being sensible of a darker shadow, quite outside this serene and active workshop of the human spirit, and yet somehow touching it. Scientific progress, which has so profoundly altered both the material and spiritual quality of our civilization, is not the sole root of its present crisis. But few men can be doubtful of its decisive part" (21).

As a testament to the place of science in the twentieth century, Oppenheimer's introduction also serves as an expression of the attitudes regarding science in the 1950s. Two items are of particular interest here. The first is his acknowledgment that science has profoundly affected not only the material changes the American cultural landscape has undergone but also the *spiritual* nature of our social consciousness. One might view these as complementary poles—whereas improvements in the state of human existence had undoubtedly come about as a result of science, so too had this reality changed our views of our relationship to the earth and the heavens. Bertrand Russell, in his 1951 collection of lectures, *Science and Society,* describes the effect of science upon our view of humanity's place in the universe this way: "[Scientific progress] has at once degraded and exalted [mankind]. It has degraded him from the standpoint of contemplation, and exalted him from that of action" (12). A fast-paced, industrial society does not appear to have the leisure to think reflectively as it might once have done, but this same society has made enormous strides in production, technology, engineering, and consumption. The opposition inherent in having the ability to create but not the time to consider the implications of one's labor is perhaps central to the Cold War condition, which may be defined by a shift in emphases creating confusions about a collective (and even individual) role in our changing social landscape, a struggle that resulted in some of our most errant political actions. On an even larger scale, where science had removed us from our position of centrality in the universe, it had made us more answerable to our own actions, our own influence on the world, our own destiny. In the face of advances in astronomy, physics, and biology, it had become increasingly difficult to see ourselves as the "purpose" behind the universe. Whereas Dante's cosmos was small, tidy, and ultimately driven by human affairs, human conduct, and human interaction with God, new scientific knowledge makes such a perspective quite implausible. How is it that, as only one tiny sphere circling one modest star amongst billions of stars in a galaxy amongst billions of other galaxies, we can hold onto the notion that we are somehow the final end of the universe?

This cosmic decentralization of humanity was hardly new in 1950, yet Oppenheimer has articulated the same degradation and exaltation in his introduction, suggesting that we are going to find it increasingly difficult to turn to God as a source for moral inspiration when we have so successfully removed him from our proximity. Scientists may not be gods, but they (and we) are suddenly saddled with godlike responsibilities, not the least of which is the control of our own future and the preservation of our minuscule corner of the universe. The real emphasis seems to be not on the relationship with science and the prospect of godhood, however, but on science tainted by the interference of domestic and foreign policy and international politics. Oppenheimer, recognizing the new spiritualism that science engenders, also recognizes the potential for corruption that the realization that we are our own gods brings with it. If power corrupts, then science represents that absolute power that may corrupt absolutely, and Oppenheimer, despite his upbeat tone in much of this introduction, cannot help but give us a cautionary stricture: "The need for the practical fruits of science is worldwide, as universal as man's striving to improve his lot on earth. The community of science is a limited but worthy prototype for that tolerant, open, open-minded community of men which alone can maintain the progress of civilization, which alone can contribute in these critical times to fulfilling the aspirations of mankind" (23).

Whether Oppenheimer believes that science, left to its own devices, can construct the tolerant, open-minded community that is necessary for the productive, benevolent advances science promises is a question only he can fully answer. However, it is clear that behind the carefully phrased warning there is a sense that the critical nature of the times in which he lived is truly reliant not on science per se, but on the ethical, mindful, thoughtful pursuit and application of scientific knowledge. He has already witnessed, firsthand, the destructive potential of science's latest technological terror, and he concedes, implicitly, that so much of science is reliant on the political and social context in which it is cultivated. It is interesting to me that Oppenheimer nowhere in his introduction explicitly mentions the atomic bomb, even though it is the latest culmination of scientific knowledge, awesome in its own right, but so charged with moral uncertainty (in his own mind as well as the nation's, which, I will later show, share many misgivings) and political pitfalls that he is forced to refer to the advancement of science using idealistically vague gestures. This odd amalgam of doubt, grandiosity, practicality, ethics, and spirituality defines the early Cold War attitude regarding science, and Oppenheimer's introduction is, in many ways, a definitive statement of the promise and uncertainty that science offered.

The second interesting feature of Oppenheimer's introduction is his hope (though I would interpret this as *appeal*) that people who are not scientists can reach agreement and understanding regarding the advances of science (through what, besides the "greatest possible contribution to human welfare," such agreement and understanding entails he does not say). This can be read at least two ways: that the nonscientist is the educated layperson interested in the outcomes of scientific labor; and that the nonscientist is the statesman or politician who so often decides the path of scientific enterprises. It is imperative, he suggests, that in order for science to reap the fullest possible harvest of which it is capable, everyone not attached to the scientific community must have a working knowledge of the scientific project, however that might manifest itself, lest science be used in ignorance and ambition—the results of which could be disastrous.

The amount of money and effort expended to educate the public and to recruit new talent into scientific fields is evident in governmental spending toward this end. In the May 1953 issue, *Scientific American* reports that the American Council on Education released advice to both the government and to U.S. colleges recommending that too much emphasis is being placed on the practical results of technological development and not enough on the more humanistic dimension of science. According to *Scientific American,*

> the report, prepared by a special Committee on Institutional Research Policy, notes that the Government is now spending $150 million a year to support academic research. Most of this money is allocated to "hardware" projects—those from which quick practical results are expected. They are frequently in fields where security considerations prevent the publication of results. The projects often pay salaries far higher than the normal academic scale. As a result, says the report, academic research is now leaning too heavily toward applied science, and there is reason to fear that students are not being trained along lines that will produce a new generation of creative scientists. (53)

The piece goes on to say that the committee recommends that colleges and universities reject all classified projects unless there is a "compelling emergency" and that these same institutions should keep their own philosophies and goals in mind when accepting governmental funding. The fear was, apparently, that when institutions become too reliant on the government for financial support, the spirit of academic freedom and the ethical application of scientific discovery are in jeopardy. Education, in this case, is meant in the classical sense of free exchange of ideas and creative development of scientific innovations, not in the limited, short-sighted interest of a governmental agenda. The reality, of course, was that many institutions not capable of sustaining themselves without governmental support would be happy to adopt the party line if it meant increased funding, new facilities, and job security. This was at the heart of the Council on Education's fear: the carrot dangled in front of universities was so tempting that the humanistic element of education would be lost, transforming all research scientists and students into mere pawns for the cause.

The government was, in fact, holding the purse strings in a very uncomfortable way. In the August 1953 issue, *Scientific American* reported that "more than half of all scientific research and development in the U.S. is now sponsored and paid for by the Federal government" (40). The "Science and the Citizen" piece, entitled "Where the Money Goes," reported that "of the $1.9 billion disbursed for research and development by the Government (mainly the Defense Department and Atomic Energy Commission) last year, $338 million went to non-profit institutions, mostly large universities" (40). This money went to research centers such as the Los Alamos Scientific Laboratory and the Johns Hopkins Applied Physics Laboratory. Because so much money was being allocated to these projects, the National Science Foundation, according to *Scientific American*, raised questions such as these: "Should universities do so much military research? Should the Federal funds be spread among more institutions? Is too much being spent in the physical sciences and too little in the biological and social?" (40). There was clearly concern amongst scientists that their contributions to new knowledge were too politically charged and that the financial motivator that the government was providing would not turn out independently thinking scientists, but ideologically indifferent workers seeking only to remain funded and employed, regardless of the political, social, and environmental ramifications. Universities were not the only agencies vying for governmental favors. Private corporations like Lockheed, Boeing, Melpar, Bendix-Friez, and scores of other technologically based companies were staging an all-out recruitment campaign. A Lockheed advertisement in the July 1953 *Scientific American* listed immediate openings for aerodynamicists, airplane specifications engineers, ballistics engineers, dynamicists, flight test engineers, flutter and vibration engineers, research engineers, instrumentation engineers, research engineers, scientists for systems analysis and military operations research, servomechanism engineers, stress engineers, structures engineers, thermodynamicists, and weight engineers (85). These skills would be applied to the research and development of fighters, bombers, trainers, cargo transports, radar search planes, and luxury liners. Lockheed even listed ads recruiting engineers in all fields who would be trained at the company's expense (*Scientific American*, March 1952, 37). Boeing ran similar advertisements, as did Chance-Vought, Grumman, and McDonnell Douglas—all aircraft-design companies. Materials companies were also demanding engineers and scientists for their governmentally sponsored research. Melpar Research Laboratories announced training programs for engineers and physicists in fields such as mechanical engineering, servo engineering, operations research analysis, electrical engineering, chemical engineering, and general physics. Clearly, the demand for scientists was great, and such opportunities were available largely because the government required—and was willing to fund—developments that would keep its military state-of-the-art.

So Oppenheimer's proclamation that science would provide hope for America's future was represented in the demand for scientists and the perceived need to inform the general public. It was, in short, a simple articulation of the Cold War scientific climate. But he also wanted to appeal to the humanistic aspect of scientific pursuits because, if we did not enter the scientific realm with our eyes open, we were doomed to make catastrophic mistakes. Without enlisting the language of panic or alarmism, Oppenheimer skillfully constructed a message of anticipatory glory and grave counsel in a single, sweeping statement. This Cold War schizophrenia is also definitive. We placed so much faith on the shoulders of science that it would be imprudent to abandon it because it had calamitous potential. It had won us a long and bloody war, yet in that war it had also given us advances in energy, aviation, agriculture, and medicine. The anxiety surrounding the prospect of scientific progress is at once bizarre and understandable. Who can argue that people in developed countries live longer, better, more fulfilling lives in large part because of the advances in science? Did anyone really wish to return to the days of subsistence living, where one family, if it was lucky, produced only enough to keep itself alive? While such a prospect did carry with it certain Hobbesian freedoms, it was hardly an appealing liberty. Science had fed, clothed, housed—and slaughtered—millions. So ambivalent was the pledge of science in 1950 that it surely must have seemed like a technological pit bull: properly fed and treated, it would be a friend for life; abused, it was capable of destroying its owner.

Such a statement should not be viewed as hyperbolic. There truly was a fear and excitement in this most recent Age of Science. Oppenheimer provides an overview of the attitudes that drove the Cold War mind-set, casting it as that which had been done, that which could be done, and that which must be done in the name of science. As inheritors of this scientific legacy, we must understand the level of uncertainty surrounding these critical times. Without the appropriate scientific awareness of the public at large and of the politicians who governed them, science could easily fall prey to ignorance and reactionism. The push to educate the common person in the ways of science took a number of forms as a result. Whereas a man like Oppenheimer would prefer to have a population of scientifically literate Americans, he was also aware that large-scale fluency in such esoteric fields as theoretical physics was impossible. Methods of a more rhetorical nature arose and were common in this campaign to educate the public in the ways of science.

Scientific American had an entire section dedicated to the monthly pursuit of this end called "Science and the Citizen." Reports by the Atomic Energy Commission (of which Oppenheimer was chairman of the General Advisory Committee in 1950), civil defense, artificial intelligence, biology, peaceful uses of atomic energy, and the latest concerns surrounding the development of thermonuclear weapons were all topics exhibited in this section, designed not for the hard-core scientist, but for the educated and interested layperson. The push to inform the laity of developments in science, however, was not without its problems in relation to the issue of national security, especially as it regarded new developments in atomic weaponry. The ambivalence between a perceived need to let the public know what was happening in the world of thermonuclear weapons (and other key scientific developments) and the governmental paranoia about this information falling into the wrong hands is reflected in the "Science and the Citizen" section of the May 1950 issue of *Scientific American,* in an article titled "Concerning H-Bomb Reactions." It begins:

> The first major extension of censorship over scientific information in the U.S. since the end of the war was instituted last month. The Atomic Energy Commission directed its employees and consultants to cease public discussion of thermonuclear reactions in relation to the projected hydrogen bomb. The Commission explained that its request applied to unclassified as well as classified information. It was addressed to all persons now or recently associated with the atomic energy project, which would include most of the atomic physicists of the U.S. (26)

According to the article, this policy had been applied to a report entitled "The Hydrogen Bomb: II" written for *Scientific American* and the *Bulletin of the Atomic Scientists* by Hans A. Bethe, a professor of physics at Cornell. *Scientific American* was forced to stop its presses, and an AEC officer oversaw the destruction of the type and the three thousand copies of the magazine that had already been printed. *Scientific American's* publisher, Gerard Piel, had this to say about the AEC's action:

> We consider that the Commission's action with regard to the Bethe article and the sweeping subsequent prohibition issued to the nation's atomic scientists raises the question of whether the Commission is thus suppressing information *which the American people need in order to form intelligent judgments on this major problem.* While there are certainly areas of information which must be protected for reasons of national security, there is a very large area of technical information in the public domain *which is essential to adequate public participation in the development of national policy,* and on which the American people are entitled to be informed by such recognized authorities as Dr. Bethe. (26; emphasis mine)

Here we see an example of the patriotic and intellectual conflict that made the Cold War such a volatile set of circumstances. On the one hand, we have the chairman of the Advisory Committee of the AEC, Robert Oppenheimer, emphasizing the importance of an informed public on matters of science, but the same commission for which he works, a governmental agency, is censoring information that is vital to that understanding. We have *Scientific American,* for which Oppenheimer wrote, dedicated to the pursuit of a scientifically informed public, but superseded by governmental strictures and, in a rather anti-American move, seizing and destroying materials that would have been published—an act in direct and unambiguous violation of the First Amendment. On the other hand, the issue of national security was considered critical enough by the government that such basic rights could be disregarded in the interest of preserving the very freedoms that were being violated. But rather than resort to liberal overstatement, it should be noted that *Scientific American* was no irresponsible tabloid; it had the respect and support of the scientific *and* the lay community. There is no evidence, in the issues that I have studied during the time frame of 1950-58, to suggest that *Scientific American* was disingenuous in its assertion that it *was* writing in the interest of informing the public about scientific developments, not because it was a good way to boost publication (though that would certainly be a happy residual effect), but because its editors held faithfully to the premise that an informed public was more likely to make rational, correct decisions regarding the application of science and technology in matters of public policy.

This troubling situation raises a number of philosophical questions about the emergence of governmental processes when confronted with the need to develop strategies for dealing with technical information that was floating around the domestic sphere. Of more direct interest to us here, however, is the statement of the problem itself and how this defines the rhetorical approach adopted by scientists and their media spokespersons to ensure that the public did have the information it needed to understand the complexity of this scientific quandary. This chapter engages the question of just how to articulate the public dilemma I have described above, one which put considerable strain on the moral and rational makeup of scientists and their advocates preoccupied by the question of how, when, and to what extent the public should be educated about scientific activities. My primary sources for this chapter are *Scientific American, Popular Science,* and *Life* between 1950 and 1960 in an effort to establish the rhetorical and ideological strategies used when communicating with the public about scientific matters in general and the atomic bomb issue in particular. Also of interest here, as throughout this book, are the cultural associations derived from religious and philosophical sources when expressing scientific concepts and options to the public during these same years.

Profile of the Scientist: The Iconographic Mythos

Before discussing the methods used to educate the population on the ways of modern science, it is useful to understand some of the assumptions made about the characteristics of the typical scientist, because it is through these assumptions that scientists gained much of their authority. Through the continuing knowledge of nuclear power, scientific writers were able to capitalize on two fundamental attitudes regarding science in general and nuclear energy in particular: science (and the scientists who apparently controlled it) had the potential either to save humankind or to destroy it. Both of these ideals were pregnant with the images that had evolved through a selective understanding of what the scientist did and was: some saw scientists as sequestered eccentrics who, in their white coats and sheltered laboratories, impiously meddled with forces of nature that they only vaguely understood; others viewed scientists as deitific guardians of knowledge and truth; still others saw them as megalomaniacs who retained a perverse need to control the world through their esoteric knowledge—what became the mad-scientist icon. All of these images held one common feature: scientists had a unique knowledge of the workings of the world, one that teetered precariously on the boundary between good and evil. In short, scientists could be either benevolent or malevolent, but the dividing line between these extremes was exceedingly thin.

These polar views of science may be inherited from a religious tradition that necessarily received new forms of knowledge in terms of a Christian counterpart; hence, the tendency of the American public to see scientists as either devils or saviors, a product of the residual culture making its impression on the dominant one. Kenneth Burke, as I have mentioned, has described science as an ideology that shares many important features with religion. Burke's point, of course, is not that science and religion share methodologies or even worldviews, but that they have essentially the same end in mind: to understand the human contingency in the overall scheme of the cosmos. We see a more typical stock response, however, in the tendency to cast the scientist in the simple bipolar good/evil dichotomy, a world of black and white that informed public consciousness on many social issues in the 1940s and 1950s. Just as in the communism/democracy, rightist/leftist, us/them extreme that governed political awareness in the public eye, the good/evil Christian formula of scientific worth was well established as a touchstone for judging scientific enterprises. Selling the positive side of science became a high priority.

In order to do this, it was necessary to humanize the scientist, to overcome the stereotypes and show that the scientist was not simply an ego-driven kook with a godhood complex. Even more pressing for public welfare and national security was the need to recruit new scientists, and this was seen as a national crisis, one that meant the difference between winning the Cold War and losing our democratic way of life. In the November 1952 issue of *Scientific American,* Anne Roe, a clinical psychologist and the wife of paleontologist George Gaylord Simpson, wrote an article entitled "A Psychologist Examines 64 Eminent Scientists," which was subtitled "The present shortage of qualified scientific workers raises the question of how they are made" (21). We might here make the distinction between "scientists" and "scientific workers," a distinction which is not expressly made in the article. As we well know, not everyone has the intellectual rigor and mental acuity to be a scientist, even though there are many and varied fields that demand a range of abilities and aptitudes. But implicit in the term *scientific workers* is a careful delineation between the research scientist and the worker who merely possesses some scientific skills and training. Such workers are especially useful in industries that have the military as their primary contractor: aircraft factories and shipyards and all the adjacent industries that supply materials needed for these operations to function, for example. Without qualified scientific workers, such industries could not hope to keep pace with the necessary output of military equipment. In short, where the scientist proper is responsible for the development and research of new methods and technology, the scientific worker is needed to maintain and manipulate this technology, as well as fill on demand orders for parts, machinery, and technological industry already in place.

The urgency of this need *is*, however, expressly stated in the article. Roe contextualizes her study by asking "What elements enter into the making of a scientist?" and, more tellingly of the preconceived notion of what a scientist is and does, "Are there special qualities of personality, mind, intelligence, background, or upbringing that mark a person for this calling?" (21). These questions are clinical on their face, but they also function to address the assumption that the scientist is brilliant, has opportunities that others may not, and enjoys a natural talent for the more esoteric levels of mathematics and a capacity for theoretical imagination. But Roe quickly switches to the practical interest of her study, "because the recruitment of qualified young people into science is a growing problem in our society. Where and how shall we find them?" (21). In keeping with the general mission of *Scientific American* to inform the nonscientist about scientific issues and developments, Roe is using her study to determine the profile of the scientist for the implied purpose of encouraging others who may feel they possess a similar profile to enter scientific fields.

Her findings are, in many ways, what one might expect, but her methods are also of interest. She examined sixty-four "eminent men" (note that the findings will apply only to male scientists; there are no women at all taking part in the study, and their status as eminent suggests that these men are the icons, the models toward which all other scientists should aspire): twenty biologists, twenty-two physicists, and twenty-two social scientists. The physicists were placed into one of two camps: the theoretical and the experimental. The only social scientists used were either psychologists or anthropologists. There was no mention of the types of biologists used. She is, in fact, exasperatingly vague about much of the study, using such *un*scientific phrases as "a high percentage," "a number of the individuals," "many," and "several." She notes that the study "has developed a great deal of material," but that in the *Scientific American* article, "it is possible only to recapitulate the high points" (21-22). While this is undoubtedly true, the reading of her audience is a bit patronizing. She writes the entire article in remarkably simple language, as if she has only two modes of articulation: highly technical or condescendingly easy. It is as if the scientist Roe (scientists have since questioned just how much of a science psychology is, and for good reason) cannot shed the clinical part of her personality, the assumption that the layperson is incapable of understanding anything but the most straightforward representation of scientific ideas.

This, in itself, helped perpetuate the mythos surrounding the scientist—the idea that scientists are better, smarter, and more sophisticated than the rest of us. Roe's study does not do much to dispel this myth, which is ironic, since one of her ostensible aims is to help recruit new scientists but is, perhaps, the reason for the distinction between "scientist" and "scientific worker." Using three different tests—an intelligence test (she does not say which one because, as she later reveals, all existing tests were too undemanding for the scientists), the Thematic Apperception Test, and the Rorschach inkblot test—she provides a profile of the scientist that is in some ways obvious, in some ways surprising, but that is ultimately overgeneralized and not very useful from a clinical perspective. Take this description of the "'average' eminent scientist":

> He was the first-born child of a middle-class family, the son of a professional man. He is likely to have been a sickly child or to have lost a parent at an early age. He has a very high I.Q. and in boyhood began to do a great deal of reading. He tended to feel lonely and "different" and to be shy and aloof from his classmates. He had only a moderate interest in girls and did not begin dating them until college. He married late (at 27), has two children and finds security in family life; his marriage is more stable than the average. Not until his junior or senior year in college did he decide on his vocation as a scientist. What decided him (almost invariably) was a college project in which he had occasion to do some independent research—to find out things for himself. Once he discovered the pleasures of this kind of work, he never turned back. He is completely satisfied with his chosen vocation. . . . He works hard and devotedly in his laboratory, often seven days a week. He says his work is his life, and he has few recreations, those being restricted to fishing, sailing, walking, or some other individualistic activity. The movies bore him. He avoids social affairs and political activity, and religion plays no part in his life or thinking. Better than any other interest or activity, scientific research seems to meet the inner need of his nature. (22)

The rhetorical impact of such a characterization on the general reader must have been discouraging in some respects, and the profile belies the claimed purpose: to recruit new scientists into the legions of the scientific community. The picture of the average eminent scientist carries with it some fairly specialized requirements: antisocial tendencies, self-removal from popular culture and interests, an apolitical disposition, an overwhelming work drive (and an attitude that one's work is one's life), and a hesitant, practical approach to relationships.

Even more interesting is the claim that "religion plays no part in his life or thinking," by which she undoubtedly means *institutionalized* religion. While the scientist, in general, may often shun formal religious institutions, the idea that spirituality plays "no part in his life and thinking" seems unlikely. The world's greatest scientist, Albert Einstein, was highly motivated by spiritual questions—questions that had cosmic significance and were driven by a need to answer the most basic religious questions: Who are we? What is our role in the universe? How did the universe begin? These questions, though pursued using science as a method, are highly spiritual and highly motivated by the idea that there is something beyond us and greater than ourselves—a basic impulse of the religious quest. The traditional assumption that science and religion are at odds is one implication of Roe's blanket statement. This was a typical outlook in the 1950s, one that has been revised somewhat today. The idea that science and religion are diametrically opposed perspectives is one of the stranger myths in the history of science. It stems from the historical events that have been written to portray religious leaders as ignorant, superstitious fools more concerned with protecting their own locus of power than with the pursuit of knowledge, while science was a lighthouse in the fog of theological myth, doctrine, and corruption. In the rhetoric of science of the 1950s, this was the portrait of science that had been inherited. Roe's statement—which ostensibly is an objective report of the scientists' own statements—buttresses the assumption that scientists were, in effect, above the dogma of religion as a social institution. It does not sufficiently account, however, for the humanistic motives that underpin many scientific enterprises.

A related myth that Roe helps to perpetuate is the notion that scientists pursue knowledge strictly for knowledge's sake, that there are no emotive drives, no preferences, no desired outcomes to their experiments or theories. These suggestions are reflected in her overall profile, especially when she discusses research as the drive in these scientists' lives—not the means to an end, but the end itself. In referring to early scientific training, for example, she states "that research experience is so often decisive is a fact of very considerable importance in educational practice," and that "the final decision to become a scientist is the discovery of the joys of

research" (25). Without overinterpreting these statements, it is important
to notice that they subtly reflect an image of the scientist as the sequestered workaholic, defined only by his desire to discover and identified
only by the results he produces. Research, though objective and single-
minded, is a joy—the only thing that gives the scientist's existence purpose
and direction. Related to this is the observation that these scientists did
not show an early interest in women, married late (age twenty-seven, on
average), and had family relationships that provided stability. The reader
can here only fill in the gaps, speculating that the reason for the scientist's
late interest in the opposite sex is a result of his preoccupation with scientific research, another stereotypical image of the scientist in the long
white laboratory coat, hunched over his test tubes while his beautiful as-
sistant tries futilely to seduce him. He cannot be bothered with sex; there
are important and *rational* discoveries to be made. However, when the scientist did settle on a mate, it was done in much the same way as he conducted research: logically, rationally, and permanently.

No intelligence test devised, in Roe's words, was "sufficiently difficult
for these eminent scientists," and the Educational Testing Service therefore
devised a special IQ test. Even then, neither the experimental nor the theoretical physicists would agree to take the test on the grounds that it was
too easy (23). Intelligence, of course, is a key assumption regarding men of
science; stupid scientists are not likely to reach the status of eminence. But
the assumption reinforces the notion that high intelligence is a key to success as a scientist, a logical preconception that sets scientists apart from
the general population, many of whom find upper-level mathematics and
theoretical physics baffling and unfathomable. The esoteric nature of scientific discourse compounded the already difficult task of educating the
laity on the purpose, methods, and foundations behind scientific ideas. Scientists were seen as a cut above, an elite community of men who understood what the rest of us couldn't, and this added to the myth of the
scientist as guru of the mysteries of the universe. The language of physics
is mathematics, a tongue that most of us are only superficially versed in.
To be fluent in it does take a high degree of intelligence, and like any other
restricted language, it also serves as a measure of acceptance into a specialized community. It was impressive that a handful of scientists would
even attempt to make the public understand the tenets behind such mysteries as nuclear physics. Physics and math being what they were neces-

sitated another method of education—one that relied on master rhetorical tropes such as metaphor, synecdoche, irony, and metonymy.

The result of this portrait feeds what I like to call the iconographic mythos of the Cold War scientist. The icon is recognizable as both a picture and an idea: the picture of the scientist, while not entirely removed from the long white lab coat and test tubes of the past, was fatherly, almost deific. The image is one of a calm, wise, concerned, patriarchal personage. The scientist was therefore trustworthy, even priestlike, in demeanor, and as a community, we looked to him (the image was male, naturally) for careful, deliberate, informed guidance in areas of knowledge that the rest of us only feared and recoiled from. The scientist was the carrier of vital knowledge about life, and, like his children, we looked to him for answers and for comfort. While we could pray to our God for this same comfort, the scientist seemed more accessible, and while he and what he did seemed mysterious, there existed an illusory approachability. The humanizing effort by psychologists like Roe, ironically, had in many ways an opposite effect because it made the scientist appear exceptional—like a prophet chosen from the masses of ordinary people. The icon was also recognizable as an idea, and the idea it projected was security in innovation—technological advancement in the name of civil protectionism. No other American icon has projected the idea that knowledge is power to the extent of the Cold War scientist, and in this case, the scientist ostensibly had the power to both preserve and destroy. Given that this power appeared to be in the hands of the scientists alone, at least in any practical sense, it is little wonder that many scientists enjoyed idolization in the public imagination. There is a certain scriptural ancientness in this idea, and it taps into the American psyche during the Cold War in a tangible and unique way: since the earliest religions, the idea of a deity that was capable of both creating and destroying has dominated our myths and legends. Here, the deity was not conscripted to the nether worlds of sacred texts; it was brought to life through a very familiar but extraordinary mechanism—the human mind.

I don't want to insist that the American public saw scientists as gods, per se. The rhetorical impact was subtler than that. The iconographic mythos that enveloped the American scientist was a complicated blend of instinctual reaction and culturized rationalization, rooted in a social network of images, ideas, and expectations and validated by the scientific profiling of psychologists like Roe and the careful stylizing of scientific public relations. A pragmatic social structure demands pragmatic explanations and approaches from its heroes. Scientists fit the bill. Such a social structure also demands a pragmatic methodology for confronting its problems and for assembling solutions. Science was called upon. But we retained enough of our earlier, classical, humanistic, and religious associations that we conveniently drew upon these images and ideas from the repository of our cultural heritage to construct an icon in the scientist that was both practical and idealized, both tangible and spiritual, both approachable and aloof. Scientists were the obvious transitional icon for a society caught between the ideas that dominated the old and new generations.

Roe's picture of the "average eminent scientist," then, draws on old images, traditional assumptions, and, oddly, psychological profiling to give us a statuesque illustration of the men who had a great impact on the future of all of us. All of the characteristics she describes conspire to create the iconographic mythos of the Cold War scientist. Her depiction serves to create a visual association in the mind of the average American, one that was actually already present but was legitimated through the scientific process of psychological profiling. This iconographic portrait was a necessary rhetorical stage in the evolution of the public/scientific relationship. Without an unequivocal, stable understanding of what scientists were, it would be difficult to rally faith in their ability to resolve the pressing problems that faced the nation. By constructing a steadfast image of the scientist, science in the public consciousness could be trusted—it was as constant, predictable, and responsible as the people who practiced it. This iconographic mythos of the scientist, together with attributes that will be discussed below, helped forge a status and a confidence in science that would be of vital importance if the public were to be expected to understand the advances, the troubles, and the answers that were the communal imperative of the day.

The Popular Press, the Popular Scientist, and the Solubility Ethos

At the dawn of World War II, scientists were questioning the moral, ethical, and social effects of science. Predicting the mechanized warfare that would define the Hitlerian blitzkrieg and, in fact, all of the military operations of World War II, scientists were already taking steps to organize themselves around the ethical application of science. Even as early as 1933, emerging from a depression that had devastated not only the United States but much of Europe as well, and with the imminent rise of Hitler to power in Germany, many scientists were concerned that war was on the horizon—a war that would rely on science and technology more than any other conflict in human history. Also, the rapid progress of science had many people, citizens and scientists alike, worrying about the crisis of overproduction and the uneven distribution of wealth that was a recipe for global conflict. The world was a boiling cauldron ready to explode. By the late 1930s, the situation was critical. In the East, vast areas of Asia were being overrun by the Japanese; in the West, Germany was moving almost unchallenged through Czechoslovakia and Poland. Science and technology had much to do with the crisis: because, many felt, wisdom was unable to keep pace with technology, war was inevitable. According to E. H. S. Burhop, a professor of physics at University College, London, and a fellow of the Royal Society, "Hitlerism was itself a symptom of the failure of Western society to build an economic system capable of making the fullest use of a science and technology developing at an accelerated pace" (33).

Even before this, some eighty scientists had formed the Cambridge Anti-war Group in 1934 to address the concerns of a war they felt would irresponsibly use the benefits of science for malevolent ends. This organization rented advertising space to warn against such dangers and to speak out against Fascism, and they also held demonstrations at airfields in England and distributed leaflets in order to make the public more aware of the rising tide of war and the disastrous effects it would have if technology was a primary tool of aggression (33-34). This social and political dimension of science had precedent in such groups and would take on new life after the bombing of Hiroshima. An interesting comment by Burhop, who was one of thousands of scientists and technicians working on the Manhattan Project, reveals an odd sense of catharsis at the bombing of Hiroshima: "I am ashamed to confess that news of the attack on Hiroshima came almost as a relief from a well-nigh unbearable tension. At last the secret was out. We could return to our homes without carrying the guilty knowledge for the rest of our lives," as if this event had somehow closed the chapter on atomic warfare and could be

conscribed to the archives of history without further concern (35). Such
naïveté is surprising from a scientist, but apparently Burhop was not
alone in his feeling. Of course, his sense of relief dissolved when it became clear that this new weapon meant new adversaries. In Burhop's
words: "The world would never be the same again. Already even more
terrible weapons, such as the hydrogen bomb, seemed possible. For the
first time in human history, it was not a question of an individual or a
family, or a tribe, or even a nation being in peril. The future of the whole
of humanity was threatened. Unless the pattern of wars, that had
seemed an almost inevitable part of human development for thousands
of years, could be brought to an end, it seemed likely that our own generation would be the last" (35-36).

Interestingly, the development of atomic weapons did not increase
scientists' influence on how they should be used; scientists "had been la-
bouring under a profound illusion in supposing that this would give them
an effective voice in the use of these weapons. Scientists were still regarded as 'backroom boys,' to supply the ideas and the new gadgets, but
to be kept in their place" (36). Not unreasonably, many scientists found
this relationship unacceptable. To counter the cavalier disregard of the
scientific political voice, a group of Chicago scientists released the Frank
Report, a pre-Hiroshima document that predicted the clash between wartime allies and outlined the effects of atomic weapons, suggesting that the
device should not be used on a real target, but detonated in an uninhab-
ited area for the benefit of the Japanese so they could witness firsthand the
destructive effects without being victims of it. When this report went un-
heeded, scientists involved with the Manhattan project began to question
where their real allegiance lay—to the government funding their project or
to the citizens of both the United States and Japan. This led to some conscientious soul searching: "As never before, scientists—particularly American scientists—became conscious of the role of science on society. The
ethical dilemma had been brought home to many of them with great
force. An unprecedented wave of political action among American scientists was initiated. Pressure groups of scientists began to descend on
Washington. Many who previously would never have dreamed of political action became active propagandists for a policy on nuclear energy of
imagination and scope worthy of the project" (36). One direct result of this
effort was the formation of the Federation of American Scientists, an or-

ganization whose primary goal was to forward the productive and constructive use of science in the United States. Other groups followed, like Science for Peace and the Association of Scientific Workers. In July of 1946, the World Federation of Scientific Workers was established. It had a charter and a constitution outlining the aims of the organization, the reproduction of which follows:

(a) To work for the fullest utilization of science in promoting peace and the welfare of mankind, and especially to ensure that science is applied to help solve the urgent problems of our time.

(b) To promote international cooperation in science and technology, in particular through close collaboration with the United Nations Educational, Scientific, and Cultural Organization.

(c) To encourage the international exchange of scientific knowledge and of scientific workers.

(d) To preserve and encourage the freedom and coordination of scientific work both nationally and internationally.

(e) To encourage improvements in the teaching of sciences and to spread the knowledge of science and its social implications among the peoples of all countries.

(f) To achieve a closer integration between the natural and the social sciences.

(g) To improve the professional, social, and economic status of scientific workers.

(h) To encourage scientific workers to take an active part in public affairs and to make them more conscious of, and more responsive to, the progressive forces at work within society. (37–38)

Items a, e, f, g, and h are of particular interest here. Item a reflects a self-imposed obligation for scientists to function as protectorate of civilization and to be instrumental as problem solvers, a trademark of popular scientists that I will later refer to as the solubility ethos. Item e also figures prominently in the early Cold War as a feature of the scientific enterprise since, in order for the population to make sound policy decisions on issues of a scientific or technological nature, it was of course necessary to have some knowledge of the subject. Item f reflects a desire to humanize science more, for, it is implied, only in this way can we make correct, moral, and ethical judgments regarding its use. In order to have a political and rhetorical voice, item g was necessary to establish scientists as not only technical authorities but also members of society who commanded respect by virtue of their economic and social status. This item is perhaps the most relevant articulation of the rhetorical enterprise embarked on by scientists. It suggests an acute understanding of the political workings of a capitalist country, where the wealthy have power and influence and a corresponding social role that is respected and admired by the public. The establishment of this status was key to having a rhetorical impact on the laity; without it, the stereotypical image of the eccentric scientist would prevail, and their voice would be muted. Likewise, item h explicitly acknowledges the need to have a far-reaching social dialogue with the American public. Here we see an early articulation of the educational enterprise that scientists had assigned themselves. Like item e, this statement reflects the perceived necessity of a knowledgeable constituency, one that has the intellectual tools and necessary technical background to make informed, ethical decisions. This document, more than any other I have found, clearly voices the role that scientists wanted, and would eventually receive, in Cold War society.

During the early Cold War years, many scientists were often simply behind the scenes, doing what scientists do, but other, more prominent figures such as Oppenheimer, Teller, and Dyson frequently spoke to the public in a much more high-profile manner. These scientists were not only extremely capable physicists; they were highly effective speakers and rhetoricians, giving science a public relations dimension that was relatively new. Science was fast becoming a symbol of guardianship for the state, for without it, we were destined for destruction by our enemies. Some examples of science's role as a PR and informational resource follow, since it is through this function that science and scientists become trusted sources of knowledge, ethics, and comfort for the American populous.

The relationship between science and technology and its impact on world politics was perhaps never more important than it was during the early years of the Cold War. During this time, technological advances in avionics, communication, and, of course, weapons, necessitated an informed (or, at any rate, conditioned) public, especially when dealing with matters of civil defense. The definitive 1950s image of the family huddled together in their homemade fallout shelter—complete with canned goods, distilled water, a rudimentary ventilation system, and other survival necessities—speaks to the level of anxiety that the typical household felt about their own tenuous safety in the event of an atomic war. Most readers old enough to remember the duck-and-cover films in school can, in retrospect, chuckle at the absurdity of the recommenda-tion to hide under our desks or cover the back of our heads to protect ourselves against an atomic attack. But even these so-called mental hy-giene films reflect something about governmental and community inter-est in educating the populace on the true scope of such an attack. In the early 1950s, before the implementation of intercontinental ballistic mis-siles (or ICBM's), the main concern was with early detection of long-range bombers capable of carrying an atomic payload. It was estimated that even if 80 percent of enemy planes were shot down or disabled, the remaining 20 percent would still have a devastating effect, especially in densely populated areas. As Hiroshima had taught us, the atomic bomb was a weapon best suited for targets with a high concentration of people, since, reasonably, the blast had a limited circumference and could kill people only if they were closely situated in an urban environment. While simply killing people was hardly a sophisticated military strat-egy, most defense experts assumed that the psychological effect alone could break morale and provide an easy opportunity for the enemy to stage a more conventional invasion.

At the heart of the Civil Defense initiative, at least so far as states-men, popular scientists, and the press were concerned, was providing the public with information regarding the status of the world's latest tech-nological bane, the hydrogen bomb. *Scientific American,* in an effort to fulfill this goal, ran an entire series in 1950 on the latest theoretical, dip-lomatic, and ethical developments of the creation of this new weapon, which would be measured in megatons (energy release the equivalent of millions of tons of TNT), not kilotons (thousands of tons of TNT), like the more conventional atomic weapons used on Japan. In the first article of the series, which ran in March 1950, the most obvious questions arose: should a full-scale effort be invested in creating the hydrogen bomb? What moral concerns does this raise? What would the development of such a weapon do to our position under international scrutiny? Should we pursue arms control with the USSR *prior* to its development?

Such questions are clearly incidental to the scientific questions that would be asked to actually produce the bomb, and this in itself is interesting, since it demonstrates a clear need to consider a public position on the matter. Even if the public did not have direct input into the developmental process, it was clear by the efforts advanced by *Scientific American* that the public should be aware of the ramifications of this looming prospect. *Scientific American* even goes so far as to say that the unilateral decision by Truman to have the Atomic Energy Commission commence work on the project was one that reflected "a major issue of public policy, one quite possibly involving national existence" and that the decision was made "in a fully authoritarian way" (11). Such a statement not only represents the scientific enterprise of *Scientific American* to involve average citizens in the developments of science that might affect them most directly, but also points out that these very issues should be handled in a democratic way, allowing citizens access to information and the decision-making process.

Scientific American functioned very much as a liberal interceder in these matters and made explicit statements to this effect. It is clear, for example, that Louis N. Ridenour, the author of "The Hydrogen Bomb" and a physicist who worked on radar development during World War II, finds the lack of candid discussion regarding the hydrogen bomb with the American public irresponsible. In reference to Truman's decision, he says that it was

> not without public discussion, to be sure, but without anything that could have been called informed public discussion. The public did not even know, and still does not, what the actual questions at issue were. Indeed, the matter became accessible to public discussion only because a careless reference to it was made on a television broadcast by a Senator who professes devotion to the principle of suppressing important information from the public, in the name of what he calls 'national security.' But the Senator did not reveal the nature of the issues and alternatives presented for decision. (11)

The irony, as Ridenour so astutely notices, is that the only information about the bomb released to the public was given inadvertently and under the auspice of national security. It is encouraging to see a scientist so fully dedicated to the furthering of public knowledge on matters of nuclear policy, and *Scientific American* stands alone as a publication that challenged the decisions of our top officials, especially when they were made surreptitiously, without public knowledge or consent.

Beyond this, Ridenour's stance represents not only a genuine desire to educate the public but also a push to use scientists as the didactic liaison for these efforts. The rhetorical effect is one of simple identification. Ridenour, and scientists like him, display an honest sincerity (as opposed to a feigned one) by taking up such positions, thereby shrinking the intellectual and ideological gulf between the average citizen and the concerned scientist. His expression of political acuity indicates that he is in the corner of the public, and that he believes in the necessity of candid dialogue for the nation, not just a handful of policy makers, when it comes to addressing issues that affect the whole country. One even gets the sense that Ridenour fears the outcome of Truman's directive if left to politicians (the chief executive officer was, after all, an ex-military man, though not nearly as staunchly military as his successor, Dwight D. Eisenhower) and wants to enlist a system of checks and balances that can only be monitored by an aware, interested, and informed civil body. He says, for instance, that "there is indeed a question . . . whether the technical decisions being made in Washington are being arrived at in a manner to inspire confidence that they will promote the greatest security" (14). He clearly disdains the whole notion of national security when it is used as a veil for making unilateral (and sweeping) decisions that will have a profound impact on not only individual citizens but also the entire world. He appears to view policy makers not as benevolent seekers of global peace, but as men driven by their passion for increased power. Certainly, a superbomb is indeed a big stick to wield, and the temptation of such men to develop it as a tool to bludgeon, not pacify, seems to weigh heavily on Ridenour's mind. By aligning himself with the interests of the public, and providing the background to make informed, rational decisions, Ridenour creates an ethos of the scientist as both a highly trained technician and a conscientious sentinel of the democratic process. Readers must have found this combination both comforting and appealing; the projection of the scientist becomes not only one of reassuring competence, but also one of moral and ethical integrity.

A good deal of the rest of the article deals with explaining the technological intricacies of developing a hydrogen bomb. One would expect this since Ridenour professes to inform the public on these matters. However, he returns to the adjacent questions posed in the introduction, especially whether the creation of the hydrogen bomb will have the deterring effect that its supporters claim and whether there exist other, more benign alternatives to the superbomb. He also reiterates his central point, that a democratic process is the best policy when arriving at the arms race decision. He questions, for example, the powers and exceptions we have granted the Atomic Energy Commission: "We have given the Atomic Energy Commission unusual powers, encroaching thereby on many of the principles basic to Western democracy. One gathers from recent news reports that, even after these sacrifices, we have still failed to obtain from the Commission the 'efficiency' and boldness of decision that are sometimes urged as the major benefit of totalitarianism. Our general security policy bears reexamination" (14).

I find this politicized statement of discord especially interesting for what it implies about the nature of Western democracy, and herein we have one of the more peculiar inconsistencies of the early Cold War. Having just emerged from World War II not only intact but even prospering, the arms race of this era was an expression of commitment and protectionism. America had lost half a million troops while fighting for democracy, freedom, and the American way (whatever that is, exactly). By doing so, we implicitly (and sometimes explicitly) vowed to maintain that way of life. Ironically, in order to protect what we had won during the war, many felt that we were forced into a position where we must forgo the very democratic processes we had fought and died for. The powers granted the AEC were an obvious circumvention of the public's right to involvement (i.e., representation) in decisions affecting national policy. Ridenour correctly challenges this move by suggesting that we have allowed statesmen and politicians to take control of an area that is of the utmost public concern. He says: "But of one thing we can be sure: the whole experience demonstrates once more the truth for which the outcome of World War II provided such compelling evidence—an informed democracy is the strongest and most viable political form. A government does not adequately protect its citizens by taking decisions for them that they can neither know about nor take part in. Almost certainly, we would be better off to treat the atomic energy field like any other normal part of our complicated technology, such as steel or the automotive industry" (15).

Again, the idea of the scientist as political ethicist is an important facet of the Cold War experience and an important facet of the emerging scientific ethos. It suggested that scientists were not one-dimensional in their pursuit of scientific discoveries—that they were not blind to the ethical ramifications of the technology they produced. On the contrary, they were not only deeply concerned about the effects of technology like the hydrogen bomb, but also extremely sensitive to the larger philosophical and civil questions that stemmed from this technology. Ridenour skillfully sandwiches his discussion of the technical apparatus behind the hydrogen bomb between an articulation of the social consequences and the policies and processes used to create the new weapon. In doing so, he projects an image of the scientist as an imperative civilian advocate, protecting our right to participation in the same way that the government professed to protect national security. This compelling triangular relationship, one where the scientist had an inside track on the developments of nuclear weapons and the government's methods for employing them, made the scientist an effective spokesperson and an essential collaborator in the eyes of the public. While governmental activity was dubious, at least with the ethically vocal scientist we had an ally against both the international threat of the Russians and the domestic threat of our own paternalistic government.

The second installment in the "Hydrogen Bomb" series, published in April 1950, was written by Hans A. Bethe, a theoretical physicist instrumental in the development of the first atomic weapons at the Los Alamos Science Laboratory between 1943 and 1946. At the time he wrote this piece, "The Hydrogen Bomb: II," he was a professor of physics at Cornell University. These credentials, impressive as they are on their face, are enough to lend Bethe authority on the scientific and technological details of the problems in developing the H-bomb. However, this is only part of his self-assigned task in the article. His task, he says himself, is "to clarify the many misconceptions that have crept into the discussions of the H-bomb in the daily press" and to "take up the moral issue and the meaning of the bomb in the general framework of our foreign relations" (18). In keeping with the role that the scientists in this series assigned themselves, and with popular representatives of science in general, the modus operandi was to help the general reader understand the intricacies of nuclear technology, albeit on a necessarily superficial level, so

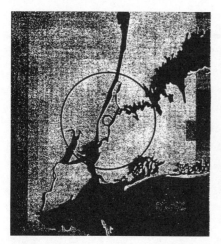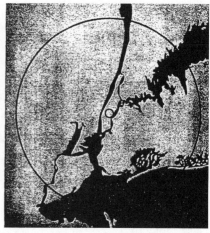

BLAST EFFECT of present and proposed atomic weapons is projected on a map of New York City and the surrounding area. A uranium bomb set off above the Scientific American *office in midtown would cause severe destruction within a radius of a mile (small circle); a hydrogen bomb 1,000 times more powerful would cause severe destruction within 10 miles (large circle).*

FLASH EFFECT of a hydrogen bomb 1,000 times more powerful than present bombs would be relatively greater than its blast effect. The Hiroshima bomb caused fatal burns at distances up to 4,000 to 5,000 feet (small circle). A hydrogen bomb would cause fatal burns at distances of 20 miles or more (large circle). The inhabitants of Chicago and its surburbs could thus be wiped out.

Figure 1. Illustration from Hans A. Bethe, "The Hydrogen Bomb: II,"
Scientific American **182.4 (April 1950).**

that the public could better understand the ethical, social, and political implication of the technology being introduced. This dual role, one in which the scientist is both technician and pedant, was a defining feature of popular science during this time frame.

I will forgo any discussion of the technical aspects of this article in favor of the equally lengthy moral discussion Bethe advances. Suffice it to say that the preparatory technical material creates a rather strange contrast to the humanistic tone Bethe adopts in the second half of the article. The technical information serves, almost incidentally, to voice the problems of H-bomb development, but primarily to show just how much more powerful and devastating an H-bomb would be compared to its uranium counterpart. Illustrated maps of the projected blast and flash radius of the two weapons demonstrate this (fig. 1).

This illustration is in some ways a fairly tame demonstration of the destructive capabilities of the H-bomb, but it does impress upon the viewer the increased range of damage such a weapon would cause. As such, it functions as an effective transition into the ethical discussion that follows. The questions the development of the H-bomb raises are numerous and intimidating. Bethe concedes that, since the decision has effectively been made for the American public—a decision he does not agree with—it is necessary to move onto the next level of moral inquiry—namely, in what light does such a decision cast the United States as a global power and the self-appointed arbiter of peace, liberty, and prosperity throughout the world? Specifically put, Bethe says the following:

> I believe the most important question is the moral one: Can we who have always insisted on morality and human decency between nations as well as inside our own country, introduce this weapon of total annihilation into the world? The usual argument, heard in the frantic week before the President's decision and frequently since, is that we are fighting against a country which denies all the human values we cherish, and that any weapon, however terrible, must be used to prevent that country and its creed from dominating the world. It is argued that it would be better for us to lose our lives than our liberty, and with this view I personally agree. But I believe that this is not the choice facing us here; I believe that in a war fought with hydrogen bombs we would lose not only many lives but all our liberties and human values as well. (21)

Bethe sees in the decision to create the H-bomb a certain moral and logical inconsistency. He worries not only that we will be viewed as hypocrites, but also that the very thing we profess to be fighting for will be lost if such a fight comes to pass. He is also deeply troubled by the idea that the United States might go down in history as the modern equivalent of Genghis Khan, and that we will be remembered not for the ideals for which we fought, but for the method we used to try to preserve them: "We believe in peace based on mutual trust. Shall we achieve it by using hydrogen bombs? Shall we convince the Russians of the value of the individual by killing millions of them?" (21). He adds that the ideologies buttressing the arguments for the bomb's development would be moot in the wake of a nuclear war, since there would be "nothing that resembled present civilization" remaining to apply such lofty ideals to. He implies here that grand ideologies are the privilege of an advanced civilization, that in the aftermath of nuclear annihilation, the only ideal to be had would be the one most basic to all life: survival of the fittest. High ideals, like high art, music, literature, culture, thrive only in a society that has a leisure class committed to the development of such things and not preoccupied with the daily annoyance of merely staying alive.

I find this an intriguing, brilliantly obvious line of argument, one that was clearly overlooked by many of the advocates for the H-bomb's development. While the theme of a new Dark Ages was one that would become popular in Cold War lore (see, for example, Walter Miller's 1959 novel, *A Canticle for Leibowitz*), few people expressly considered the absurdity of fighting for a cause that would be irrelevant under such circumstances. This level of clarity and rationalism is a feature that helped establish popular scientists' authoritative role during the early Cold War because it provided an effective antidote to the inflamed, often hysterical reasoning that dominated many of the Cold War debates. As the voice of reason, Bethe and his fellow scientists interested in the broadly applied education of the American citizen showed both intellectual prowess and an ethical consciousness. This in itself was a successful combination, for it established scientists as more than the sum of their discoveries and accomplishments; it showed our uneasy citizens that they were also *human*.

The status of science is another concern for Bethe, not so much because science stood to lose authority or dominance in the postcataclysmic world, but because people would learn to fear, distrust, and outright despise science for the evil it had foisted on them: "The destruction of cities would set technology back a hundred years or more. In a generation, even the knowledge of technology and science might disappear, because there would be no opportunity to practice them. Indeed, it is likely that technology and science, having brought such utter misery on man, would be suspected as the works of the devil, and that a new Dark Ages would begin on earth" (21). Here we see yet another layer added to the ethos of popular scientists. They were qualified to teach science to the public and guide them in matters of moral concern in order to maintain support for their enterprise; but it was equally important to act as public relations advocates for what they did as scientists. This warning is a dire one; Bethe does not want to experience a retrogressive movement in the evolution of science, and he fears that a nuclear war would have exactly that effect. The phenomenon he predicts is pragmatic and ideological. In the event of a nuclear holocaust, science would not be widely practiced for two reasons: the opportunity to conduct science would be secondary to the need to subsist, and people would fear science as an instrument of evil.

Bethe's article reveals the multifaceted image of the popular scientist in an explicit way. It is clearly written to emphasize the scientist and the scientific project as entities that are accessible and in need of moral dialogue. The ethos of the scientist that is generated through this article (and the rest in the "Hydrogen Bomb" series) makes complicated scientific processes easy to grasp for the sole purpose of constructing a large-scale ethical consciousness. In the same stroke, Bethe manages to defend scientific enterprises—even ones as dubious as the development of hydrogen weapons—by showing that scientists are not beyond the moral dictates that govern everyone else's conduct. By clearly articulating the need to reconsider the bomb's construction, he manages to undermine a myth surrounding the scientist that states, in effect, that scientists tend to be amoral and irresponsible, somehow above the humanistic concerns that the rest of us hold dear. By showing his political conscience, he further suggests that scientists are not mere pawns of a government that seems bent on circumventing the democratic process and maintaining a global power structure regardless of the consequences. He projects a confidence in the American public to make informed, intelligent decisions, and in so doing, has lent them more credence than their own government, which appears to believe that the American people cannot be trusted with such sensitive information. He has shown the public respect, in short, and by striking this chord has created a sympathetic audience, one more willing to put faith in science to do the right thing and reciprocate the respect he has shown them.

The other dominant issue, discussed more fully in the third installment to this series, was the question surrounding the H-bomb's deterrent effect and its contribution to national security. One of the primary arguments for its development was that, regardless of which action the United States took on the bomb, we could rest assured that the USSR would move forward with its construction once the technical details were ironed out, an event that seemed imminent. Knowing this, many argued, we had an obligation to national security to have the bomb first.

One widely understood U.S. policy was that it would never use atomic or thermonuclear weapons aggressively; they were needed strictly for defensive purposes. Therefore, the logic ran, we must have the H-bomb so that the Russians cannot use theirs to present an ultimatum against the United States. This effectively disarmed (so to speak) the moral question in many people's minds because it meant that the creation of the H-bomb was not a moral decision at all, but rather a defensive one. It was not the fault of the United States that we were pitted against an atheistic and amoral adversary; it was necessary, however, to defend ourselves against it. This meant that the *only* reason to build the bomb was to preserve our moral way of life—a question of national security, not of moral propriety. In fact, many went so far as to say it would be *immoral* to allow a totalitarian regime like Soviet Russia the opportunity to dominate the globe by failing to protect the world (read: the American way) with our own technology.

Hence the focus of the discussion by Robert F. Bacher in "The Hydrogen Bomb: III," published in May 1950. Bacher, like the other authors of the "Hydrogen Bomb" series, was a physicist, and he taught at the California Institute of Technology at the time this article was written. Prior to that, he had worked as a member of the Atomic Energy Commission from 1946 to 1949. Again, his experience, like the others', was central to the early policies governing the use of atomic energy and weapons. His credentials, both as a professor and as a member of the AEC, make him a leading authority on the issues surrounding atomic weapons, emphasizing again that *Scientific American* was not in the habit of using pseudoscientists when informing the American public on issues of scientific concern. Bacher's stance is somewhat less didactic than Bethe's, at least insofar as the moral issues are contemplated. He is more concerned with the degree to which the H-bomb is a viable and effective military weapon. His argument in this regard is straightforward: given the limited delivery capabilities of such a weapon, and the strategic effectiveness of simply blasting an area to smithereens, the H-bomb is in fact less effective than the existing fission bombs since they cannot be used to pinpoint specific military targets in any controllable way. In other words, using an H-bomb on military targets, or as a tool for all-out warfare, is a little like blasting a hornet's nest with a howitzer—although it will get the job done, most of the effect is simply superfluous: "From the standpoint of its military effectiveness, there seems to be little reason to attach such great significance to the hydrogen bomb. While it is a terrible weapon, its military importance seems to have been grossly overrated in the mind of the layman" (14).

Seeming less sympathetic to laypeople than Bethe, but apparently just as committed to educating them on the true nature of thermonuclear weapons, Bacher blames the politicians, the exaggerated role of H-bombs, and the Red Scare for the ignorance and misinformation of the general public. In addition, Bacher categorically denies that the H-bomb will meaningfully protect the country in the event that our enemies are determined to attack us: "Pumped full of hysteria by Red scares, aggravated by political mud-slinging, the average citizen is easily convinced that he can find some security and relief from all this in the hydrogen bomb. The most tragic part is that the hydrogen bomb will not save us and is not even a very good addition to our military potential" (14). Bacher, like Ridenour, sees public information as key to formulating efficacious nuclear policies. Here again we see the scientist as political analyst and public advocate. Without proper and accurate information, he argues, the public will continue to act on their passions and their fears, and the results of this could be serious since it would lead to a panicked citizenry overwrought with anxiety based on erroneous assumptions.

> He, too, sees good reason to maintain the American democratic system, if only because he doesn't appear to trust military men and the government to make appropriate and meaningful decisions based on the interests of the nation as a whole. He views public scrutiny of defense policies and issues to be the best weapon against hasty decision making and an authoritarian state shrouded in secrecy in the interest of national security. He further abhors the idea of citizens flippantly disregarding their basic responsibility to keep a close watch on the activities of the government, especially when these activities involve judgments so vital to domestic well-being: The U.S. has grown strong under a Constitution that wisely has laid great emphasis upon the importance of free and open discussion. Under the influence of a large number of people who have fallen for the fallacy that there is security in secrecy, and of many, I regret to say, eminent scientists, who prophesy doom just around the corner, we are dangerously close to abandoning those principles of free speech and open discussion that have made our

country great. The democratic system depends on intelligent decisions by
the electorate. Our heritage can only be carried on if the citizen has the information with which to make intelligent decisions. (15)

Bacher represents the scientist cultivated to stand as an authority on political theory—or at least a solid student of the democratic process—again
emphasizing that scientists were not creating weapons of mass destruction with wanton disregard for the ethical consequences or for seeing how
they fit into the larger framework of domestic well-being. Clearly, Bacher's
position on a democracy of, by, and for the people would have struck a
positive chord with readers and effectively caused them to question a government that excluded them from the process. The issue of free speech is
of course a direct reference to the censoring of Bethe's article, but it also
runs deeper than that. The idea that a democracy can vote or speak blindly
and still be a practicable form of government is ludicrous to Bacher, and
his argument would easily win him allies with an American public
resentful of their diminutive, if not nonexistent, participation in key decisions about the development of hydrogen weapons. Just as the icon of the
scientist was a combination of idea and image, so too was the ethos of the
scientist one of theory and practice. In this case, Bacher is not a physicist
in the strict sense of the word; rather, he is a democratic advocate who
happens to have valuable technical knowledge to impart to the American
people. He is a scientist plus: he is an accomplished practitioner of his
technological craft, and this qualifies him to be an accomplished theorizer
of the governmental system in which his craft might flourish.

    Using some of the more publicized information regarding civil de-
fense topics and the development of the hydrogen bomb, *Scientific American* frequently ran editorials and informative articles on the theory and
practice of "staying safe against the Bomb." Interestingly, the widespread
public (and scientific) concern over radioactivity—that mysterious residual plague that resulted from a detonated atomic weapon—was down-
played in *Scientific American* as alarmist. In the June 1950 issue, *Scientific
American* ran the fourth in the series of informational articles on the projected impact of a full-scale hydrogen bomb attack, which stated, "With
regard to the hazard of prolonged radioactivity, this unique characteristic
of atomic weaponry has been greatly exaggerated. As an instrument of
aggression, it is probably no more effective than chemical warfare and

less damaging than incendiary bombs. The weapons of modern war are terrible enough without adding embellishment to them" (12).

The assertion that radioactivity was in fact only an incidental annoyance in relation to the heat and blast of a hydrogen bomb led to the design of a hypothetical city in which the population was evenly distributed over a thin strip of land, making a single hit by a hydrogen bomb much less effective than if the city's population is more centralized, as is generally the case in the United States' largest metropolises.[1] The theory went like this: according to the 1940 census, on which the idea is based (despite the fact that at the time of writing the statistics were ten years out of date), ten states in the northeastern quarter of the United States contained nearly 65 million people, which equated to 43 percent of the total U.S. population. In this area were concentrated cites like New York, Boston, Philadelphia, Baltimore, Washington, D.C., and scores of smaller cities. Inspired by the notion that the United States has a vast area of undeveloped land, the author, Ralph E. Lapp (former head of nuclear physics at the Office of Naval Research and an independent consultant in atomic energy), suggests that the best way to combat this dangerous concentration of people is to disperse the cities.

This, of course, would be no small task. It would require the development of cities with a population of 50,000 or less, meaning that many of the new projected cities would have to be built. The concern over suburban sprawl was apparently not as acute as it is today, since the cities would not only be numerous but also traverse huge sections of the country, none of which, as is the whole basis of the plan, would be centrally located. The minimum distance between towns, according to Lapp, would be fifteen miles, with even larger distances being preferable. In theory, the projected cities seem like a pretty good idea. They would be laid out in a strip no wider than two miles but stretching, conceivably, for hundreds of miles from end to end. Cutting through the center of the city in a straight line would be central transit—a train system that shuttles back and forth from one end of the city to the other. The ends of the city would contain the industrial centers, while slightly inward from these areas would lie the business district and even farther in, the residential areas and some light industry. On one edge of the city would be a superhighway with perpendicular exits intersecting any area of the city the traveler wished to enter. On the other edge would be farmland. Figure 2 is a rendering of a

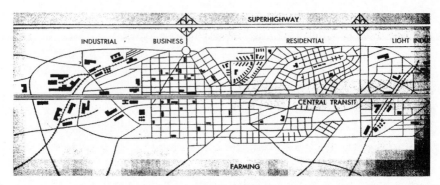

Figure 2. Illustration from Ralph E. Lapp, "The Hydrogen Bomb: IV,"
Scientific American 182.6 (June 1950).

section of the proposed city. The design is quite ingenious; but, of course, the reality of such a city never came to fruition. As practical as Lapp makes all this sound, the obvious expense and the engineering obstacles of such an undertaking made such a project quite unlikely.

Our primary interest here, however, is not such a city's viability but rather the proposal's rhetorical presentation. In many ways, the logic of Lapp's idea is inescapable: given the sheer vastness of the American landscape, and given the concentrated destructiveness of thermonuclear weapons, in Lapp's words, "space is our ally." (Russia is even larger, which makes one wonder what weapons might have been developed if such cities had become commonplace on both sides of the Pacific.) Nevertheless, the layout has apparent advantages: traffic congestion would certainly be less of a problem. With such a design, all heavy traffic would travel on the superhighway on the outskirts of the city, and the mass transit line running through the middle would alleviate much of the need to drive to work. The central transit system would be fairly easy to maintain and operate given its central position and its orientation to business, residential, and industrial sectors of the city. Most important, the residential sections of the city would be relatively safe during a nuclear attack since they are situated well away from the industrial section.

Public response to this proposal must have been a mixture of awe and relief; if nothing else, scientists and engineers were working toward a solution to civil defense problems no matter how remote or costly some ideas must have seemed. While thermonuclear weapons were being developed and manufactured by scientists, the public could at least feel secure in the faith they had for science to embrace and solve any national problem. The fact that such a proposal was being offered by a known and respected physicist only made his authority more reassuring. This was not merely a scientific writer enthusiastically showcasing the latest developments and problem-solving abilities of the scientific method; he was a top consultant proposing a logical solution to a very real threat.

Furthermore, Lapp is a convincing rhetorician. His delivery is even and rational, lending credence to an already well-established public assumption that scientists were calm and collected even when everyone else might be panicky and hysterical. Lapp's suggestion is that the weapons are frightening because they are new and the effects are overblown. Therefore, the public needs to understand the weapons better and take the necessary steps to guard against their primary mode of destruction: concentrated blast and heat. Rather than succumb to reactionary hyperbole and rash survivalist measures, the targets simply need to be rearranged so they are less appealing and attacks against them are less effective.

The aptitude many scientific spokespersons had for projecting this level of confidence and resoluteness is certainly one very significant reason for increasing reverence for science during the postwar years. It reflects an important form of scientific rhetoric occurring during this time: the solubility ethos. The solubility ethos was an effective device for reassuring a nervous public because it did two things: it created an environment where scientific knowledge was accessible (and necessary) to the layperson, and it reinforced the assumption that scientists were highly competent to recognize and solve critical problems. National priority was given to the effort to educate the public about the ways of science, and a new breed of scientific delegate emerged as a result: the popular scientist. The popular scientist was the manifestation of the solubility ethos, since it was through him (and I use the masculine pronoun deliberately) that the sweeping educational campaign could be conducted. Unlike scientists working in the scientific community, a self-contained field wherein one had to answer to other scientists and a strict code of methodology, experimentation, testing, and review by one's peers, the popular scientist really had no such checks-and-balances system to determine the validity and/or viability of his claims save for the authority that was afforded him by default through simply being a scientist. Popular scientists were usually taken at face value and lent the clout the solubility ethos created for them because the public was desperate for visible figures of authority in matters that could have a direct, profound, and physical impact on them.

Occasionally, it is true, scientists would debate publicly, usually in a forum of popular science magazines or through a series of publicized pamphlet wars. One example of this is worth mentioning, if only because it underlines the public nature of scientific debates for the purpose of educating the community. At least two professors at Columbia University, Seymour Melman and Mario Salvadori, felt that the civil defense initiatives in place might actually encourage a nuclear conflict between the Soviet Union and the United States. They had argued publicly on several occasions that instilling a false sense of security in the population made the likelihood that we might actually feel comfortable using atomic weapons greater. Melman warned that in a twelve-mile radius from the blast of a hydrogen bomb, fallout shelters would be useless. Salvadori argued that the psychological well-being created by civil defense measures might push the competing forces toward actually having a nuclear war. It is possible, then, that the very measures designed to protect us against an atomic war could actually make it more of a possibility. The rhetorical nature of this argument is intriguing. Melman and Salvadori seem to be suggesting that the rhetoric of security provided by these civil defense measures increases the likelihood that their use will be imminent. The authority extended to science is given physical proportions in fallout shelters, so that the American public sees itself as prepared for any life-threatening nuclear confrontation. Melman and Salvadori, expressing their views in the late 1950s, were proved wrong— no nuclear war occurred, and if it had, it would have been difficult to prove that civil defense measures contributed to the event. Nevertheless, these men demonstrate recognition among learned people that underscores the very nature of the Cold War: it was a conflict of language, illusion, and appearance, and if it escalated into a physical confrontation, it was the fear undergirding these appearances that would make it extremely dangerous.

The expectation cultivated in the public service and propaganda films is one of survival. Unlike previous civil defense precautions dictated in earlier wars, the civil defense of the early Cold War was one based on assumptions, as opposed to direct experience, about the nature of a nuclear conflict. The only test case for any of the procedures described was based on the explosions at Hiroshima and Nagasaki, both of which were considerably less menacing than the prospect of thermonuclear warfare that threatened the United States in the 1950s. Therefore, the films produced as informational sources of what to do in the event of a nuclear attack presented this material in far more certain terms than was actually known. In order to calm an increasingly agitated public, the films needed to construct the illusion that the authorities knew what would happen and how to protect the public. This was far from the case, however. Though scientists had been diligent in their study of the effects of radiation, fallout, and the nuclear explosion itself, much of the information given to the public was pure speculation hidden behind a veneer of authority and certainty.

Scientists like Melman and Salvadori found such films socially irresponsible and publicly dangerous since they preyed on the already inflamed anxieties, fears, and misconceptions held by the populace and created a climate of hysteria and instability. This was just one more example, they argued, of how the hype surrounding a nuclear attack would lead to erroneous policy and civic paranoia. Paranoia not only fed the fears of such an event, grounded though they may have been in the destructive potential of atomic and thermonuclear weapons, but it also increased the possibility that America would forgo its staunch defensive stance under pressure to launch a preemptive strike based on panic and misinformation. Scientists like these wished to err on the side of caution, knowing as they did that the true impact of a nuclear war was uncertain since it had never been studied firsthand except through the detonation of individual weapons. In other words, it was premature to predict the true fallout (literally and metaphorically) of such a war since it had never been witnessed. The key was to continue experimentation and study and to temper rash policymaking with a checks-and-balances system of watchfulness, diplomacy, and the careful conditioning of the American people. In this respect, the face of the scientist that had been exposed to the public was, again, one of rational forethought, careful deliberation, and prudent restraint. Since the alternative to this was panic, hysteria, and suspicion, leading ultimately to the foolish misuse of nuclear weapons and civil defense, the scientist was not only at the center

of the technological issue but also the level-headed leader figure the
country desperately needed.

Responses to *Sputnik*

In the July 1959 issue of *Popular Science* we see further examples of
science's attempts to protect the population from nuclear attack. The article "U.S. 'Space Fence' on Alert for Russian Spy Satellites" details an electronic fence that is, according to the author, "cloaked in semi-secrecy." This
fence, a direct response to Russia's recent launch of *Sputnik*, had "a deadly
serious purpose: to warn us if and when America is scrutinized by silent
space spies" (62). *Sputnik*, the first artificial satellite to successfully orbit
the earth, was launched two years earlier, on October 4, 1957, and it is generally considered to be the symbolic beginning of the so-called space race.
Its presence made scientists, politicians, and the public alike extremely
nervous, since it was assumed that such a satellite had both a practical
and an ideological purpose: practically, it suggested that the Soviets were
now capable of spying on the United States from space, a prospect that
was worrisome since no real defense against such a threat existed at the
time; ideologically, it suggested that the United States was trailing the Soviets technologically, and this insufferable prospect created some rather
hysterical reactions. Many thought that such a satellite indicated a Soviet
ability to drop bombs on us from space; others found the concept of
artificial satellites too mysterious to fathom and concocted all sorts of possible functions for a 184-pound, two-foot-diameter ball.

   Sputnik, in fact, had no practical purpose except to demonstrate that
artificial orbiting satellites were scientifically possible. Its only apparent
function was to circle the globe every ninety minutes, broadcasting a beep
at twenty and forty megahertz to make its presence known. It was, rhetorically speaking, merely a symbol of Soviet technological prowess, and
it was an irritating reminder to American scientists that they had somehow failed in their mission to stay ahead of the Soviets technologically.

   Public responses to *Sputnik* tended to zero in on the perceived inadequacies of American science. For example, interest in scientific fields increased following the launch of *Sputnik*, and we can only assume a causal
connection. There are documented writings from politicians and editori-

alists who reprimanded the educational system for its inadequate scientific training. The overall reaction seemed to be one of awe, misconception, and panic. In retrospect, it is odd so much was made of the *Sputnik* event, but the notion of space travel in any form—even such a seemingly innocuous accomplishment as sending a 184-pound ball into orbit—was until this time consigned to the annals of science fiction. With the launch of *Sputnik*, acute ambivalence pervaded American attitudes about science. The idea of an artificial satellite sent a jolt through the imagination of the American public. Like so much else that happened during the early Cold War years, it was both inspiring and frightening, and the mystery that enveloped *Sputnik* only created further speculation.

The rhetorical nature of this event manifests itself in several forms. Because the satellite was so mysterious, and because it was launched by our ideological and technological enemy, our reliance on science to understand and monitor Soviet activity seemed much more imperative. From a Soviet perspective, *Sputnik* was a rhetorical mechanism: its only apparent function was to remind the United States, with its incessant beeping, that their technology had slipped past us. Far from being shrouded in secrecy, like so many other technological advances of the Cold War, *Sputnik*'s sole purpose was to persuade America (and Russia as well) of Soviet superiority, to give us an incarnate symbol that they had made the first step toward conquering space. Space itself, as a final frontier, would be a fine prize for any superpower, even though the idea of conquering space was more than a bit grandiose, especially given the embryonic stage of *Sputnik*'s technology. However, these two rhetorical outcomes are closely linked. The idea that Russia had somehow beat us to this symbolic goal was hard to bear, and since there was little the general public or even the military could do about it, the onus of guardianship fell on the shoulders of the scientific community.

The result was the development of such early-warning systems as the space-fence. By January 31, 1958, the United States had launched its own satellite, *Explorer I* (by which time, interestingly, *Sputnik* had already fallen from orbit). It carried several scientific instruments and eventually discovered the Van Allen belt. That same year, the National Aeronautics and Space Administration (NASA) was established. The period between 1957 and the launch of *Sputnik* in 1960 represents the early years of a space race that would culminate in the first manned lunar landing in 1969. In terms of their ideological impact, the early space-race time frame saw huge changes in the American outlook of its own scientific and global role.

These events provide a way to understand and contextualize a causal relationship between the events and the corresponding rhetorical impact they had on the American public. Magazines like *Popular Science,* with the tabloid-style representation of science, did much to fan the flames of popular discord regarding technological advancement while concomitantly reinforcing public faith in science to deal with the issues such advancements raised. Using master tropes like analogy, *Popular Science* was a good source for making the casual reader understand the depths of scientific problems and the hope for scientific solutions. Consider, for example, this explanation reprinted in *Popular Science* by an army electronic engineer about the problems of using a space fence to monitor Soviet satellite activity: "[it is] like sitting on your porch and detecting a mosquito in the next county. Not only that, but telling precisely where it is, where it's going, and how fast" (64). Such analogies, while loosely accurate, accounted for the public panic because they exacerbated something the engineer probably had not anticipated: the sheer vulnerability Americans felt and the underhandedness Americans perceived as part of the Soviet strategy. How could one detect something so proportionately small? How could we arm ourselves against an unseen enemy?

Seizing upon this perception (and perhaps feeling it themselves), *Popular Science* titled the next section of the article "Naked America":

> What could such satellites see? Even from a height of more than one hundred miles, TV cameras could detect much of military value: the presence and layout of secret bases, expansion of strategic industries, major construction projects. Infrared detectors—their amazing capabilities still under security wraps—might yield more yet, telling skilled interpreters what work was going on in a specific area. Realization of all this hit the Army hard. The continental United States was stripped of privacy, the oceans no longer provided barriers to prying eyes. For the first time in American history, an enemy could watch us unawares. (64-65)

The idea of a "naked America" and "No Place to Hide," the next heading, can hardly be considered offhand phrasing, for while the titles underscore the novel, yet serious, nature of the situation, they also promote overreaction. The situation is serious but under control: scientists and engineers are on top of the problem; and while detecting a mosquito in another county is a daunting predicament, no one is better equipped and more knowledgeable than the scientist for finding a solution. This example in some ways complicates the solubility ethos of the scientist. Like the *Sputnik* launch that inspired the creation of the space-fence, this representation undermines the authority of American scientists viewed as straggling behind their Soviet counterparts, playing catch-up to a technological system that creates enormous logistical problems. On the other hand, such a situation is at the heart of the very existence of the solubility ethos in the first place: global politics dictates that the neck-and-neck race for technological superiority creates a constant push to improve existing technology. The Soviets may have beat us to the launching of an artificial satellite, but their success only created an environment in which scientists thrive. The pressure to perform—and usually succeed—in this international footrace helped continually establish and reestablish the ability of science to penetrate even the most difficult problems in the public mind.

The introduction of *Sputnik* cast the Cold War (and the Cold War scientist) in a whole new light. Where the iconographic mythos and the solubility ethos of the scientist had been watchfully cultivated and developed prior to the launch of this satellite, afterward it seemed that perhaps our scientists were not as infallible and politically attuned as the public had been led to believe. Such is the fickle nature of public judgment. Yet much of what the scientist had managed to establish was impervious to corrosion, at least in the short term. While *Sputnik* was a shock to the American public, it was also a great motivator, given the American sense of competition—that timeless drive to be number one had made the American scientist our champion, and while he may have been reeling from the last uppercut delivered by the Russians, he was by no means knocked out. The spirit of winning, and the faith in scientists despite their recent setback, established a combination of elements that eventually put a man on the moon ahead of the Russians—a far more impressive accomplishment than shooting a metal sphere into space.

In 1957, putting a man on the moon was still science fiction. Only through a combination of faith, perseverance, technical knowledge, public relations, national pride and paranoia, massive funding, governmental and public support, and scientific rhetoric during the early stages of the Cold War could such a task be realized an unimaginable twelve years later. By nurturing the image of scientists—by making them seem like the great problem solvers and by drawing on the images, fears, hopes, and anxieties surrounding science—technological progress in areas outside of the military became possible. This dual nature of the peaceful and aggressive applications of science, where one side consists of the disasters inherent in the prospect of a nuclear war and the other consists of the promise of space exploration, is a profound expression of the ambivalent and uncertain climate prevalent during the early Cold War. It is ironic indeed that many of our greatest scientific achievements occur on the heels of violent conflict or the threat of war, and never has this sad fact been more poignantly clear than it was during the fifteen years following World War II. Through the continuous cycle of build and destroy, live and die, prosper or perish, the scientist emerged as the harbinger of all things hopeful, if only because it was in science that we had placed so much confidence and devotion. It is little wonder that, in an atmosphere of impending calamity based on the technological horrors we constructed to maintain the survival of our nation, the scientist had found a prominent and vocal place in the public spotlight.

Chapter 4

New Anxieties, New Solutions, and Nonnuclear Science

Other Sources of Cold War Anxiety

Nuclear holocaust and national defense were not the only causes of apprehension for the Cold War citizen. Other problems that fell under the auspices of science included overpopulation, food shortages, and the depletion of natural resources. While these issues were by no means new to the 1950s, they had certainly been aggravated by the baby boom that had been a natural outcome of the prosperity that followed World War II. Just as the U.S. population continued to produce offspring, they simultaneously worried that there would not be enough space, food, or resources to accommodate them all. Again, problems that had largely been created by scientific advancement—industrial pollution, medical advancements that allowed people to live longer (thus creating all sorts of shortages), and urban centers that required food, shelter, and waste management for high concentrations of people—required scientific solutions. These situations created an oddly ambivalent attitude toward science. While most of the population had great faith in science and scientists to address and resolve the problems it had created, the general public did not appear to hold science accountable for the existence of the problems to begin with, an attitude that is quite understandable in the face of the progress and benefits science had created.

According to the International Data Base of the U.S. Census Bureau, the population of the United States in 1950 was 152,271,000; in 1955 it was 165,935,000; in 1960, it was 180,671,000. The increase in U.S. population was 28,400,000 in the span of a decade. This was unprecedented growth for such a short time frame, and it was a source of great consternation to the public and scientists alike. Here again we see the ambivalence that was a symbol of Cold War scientific advancement: it was impossible to deny that science had played a dominant role in the standard of living people were becoming accustomed to. The Great Depression still loomed large in many people's minds, and it was easy to enjoy the luxuries and privileges science offered when so many had done without for so long. Jobs were plentiful because of science; educational opportunities were available because of science; domestic tasks of all sorts were made quicker and more convenient because of science, leaving time to enjoy the good life created by science; people were healthier and more active in large part because of science; in general, life was easier and more prosperous because of science. Few people were willing to speak ill against the pragmatism of science when quality of life had been improved so palpably after years of hardship and war. But, lest this last statement lead the reader to believe that there was a conscious and willful resistance to science bashing, it is worth pointing out that the attitudes internalized about science did not leave the general population with a set of obvious choices about whether to accept or reject the principles, practice, and authority of science. Except for efforts of the popularizers of science discussed in chapter 3, people in large measure simply reaped the benefits of science with little active consideration of the source of their prosperity.

On the other hand, there seemed to be widespread interest in the latest developments in science, especially if the developments had an obvious personal impact. *Popular Science* ran a section entitled "I'd Like to See Them Make ..." that featured product suggestions from readers. The "Them" in question clearly referred to the engineers, inventors, and designers who worked for corporate America but who had profited from the new technology and innovations created by science. Examples included car-trunk hatches that allowed passengers to reach the trunk from inside a car; easy-out wood screws for temporary fastening; no-squeeze caulking guns; locking rifles; and illuminated dustpan brushes that could light up areas such as ovens, closets, and the undersides of furniture. The amorphous "them" would function as a reference in many contexts to the brains that drove the country's prosperity and influence—namely, the scientists. So ubiquitous had technology become, and so comfortable had people felt in using the gadgetry that had become commonplace, that the antecedent-lacking pronoun required no explanation: people simply assumed the existence of a scientific brain pool that would provide the necessary solution or make the necessary device.

Within this climate thrived a popular scientific preoccupation with anything super. Consider, for example, the poem "Superman" by John Updike, written in 1954:

> I drive my car to supermarket
> The way I take is superhigh
> A superlot is where I park it
> And Super Suds are what I buy.
> Supersalesmen sell me tonic
> Super-Tone O, for relief.
> The planes I ride are supersonic.
> In trains I like the Super Chief.
> Supercilious men and women
> Call me superficial, me!
> Who so superbly learned to swim in
> Supercolossality.
> Superphosphate-fed foods feed me
> Superservice keeps me new.
> Who would dare to supercede me
> Super-super-superwho?

Science and technology, through progressive techniques, methods, materials, and processing, had managed to superize American culture. Technology was like a magic wand that could effortlessly transform normal, run-of-the-mill objects, services, and people into superstuff. Things were super and so were people. Superman was a popular cultural icon; supermarkets thrived in a growing suburban landscape since it was no longer necessary to shop in small stores tucked away in urban streets and alleys; Eisenhower had built an interstate system in record time so people could travel with ease, efficiency, and speed along the superhighways that traversed the American countryside.[1] The country was preoccupied with the pace and products science made possible, and people had the money to spend on the new gadgetry and services that were available.

Yet underneath this veneer of carefree indulgence lurked a growing concern about the byproducts of widespread affluence: many harbored a vague concern that the very prosperity that had defined postwar life was leading to certain disaster, not only through the threat of nuclear war, but also through overpopulation, food shortages, pollution, and resource depletion. It is one of the great ironies of the postwar baby boom that the economic stability, job availability, and medical advancement that allowed society to procreate freely and safely also created the unintended consequences that such benefits engendered. And while most people chose to actively ignore these consequences, there was enough mounting concern about the palpable changes taking place in society to inspire nervousness and outright fear. It is difficult to generalize about the state of the American collective consciousness during the early Cold War, but the popular press of the time suggests that people tended to fall into one of several camps: there were those who merely enjoyed the benefits and conveniences of modern life, operating under the assumption that we were in control of any technical problem that might arise; there were those who felt that nuclear war was imminent and that they should enjoy their fleeting prosperity while they could; there were those who foresaw large-scale environmental problems and turned to science for reassurance; and there were those who took a moderate view of the situation, recognizing potential social and environmental effects, educating themselves about them, and taking steps where they could.

The society reflected in popular cultural magazines, on the radio, and in other widespread media sources was limited. The newly acquired comfort and privilege brought on by the postwar years was enjoyed largely by a well-established middle class and was, as we might expect, represented as primarily White Anglo-Saxon Protestant. While this is not universally true, it is important to note that the popular media projected this class as the norm, and they were therefore the primary target audience for most educational and informative pieces. This in turn led to the formation of dominant attitudes in this influential segment of society, making their vantage point an important one in the overall establishment of civic discourse and policy. New initiatives such as the GI bill allowed for the expansion of a capitalistic middle class, one larger than the country had ever seen before. Veterans who came from lower-income families prior to their involvement in World War II found themselves in the happy position of receiving either free or greatly reduced tuition for educational opportunities and other governmentally sponsored programs.

In fact, the GI Bill of Rights, which was signed into law in 1944, was one of the most important and influential pieces of American wartime legislation ever passed, and it had a tremendous impact on the quality of life of millions of Americans following the war. This unprecedented bill, originally presented to Congress by Franklin D. Roosevelt in 1943, included such benefits as educational supplements, life insurance, medical care, and pension and reemployment rights. In January 1944, the American legion proposed, and ultimately received, an expansion of the legislation to include both a provision for a centralized veterans' administration and one guaranteeing federal loans for homes and farms. By 1955, some 4 million veterans had used the home loan benefits, more than 5 million had received the readjustment allowance, and 7 million had taken advantage of the education and training opportunities (including 25,000 African Americans given a chance to attend college for the first time). The bill also provided millions of Americans with opportunities for education, home and business ownership, and other advantages that would have been otherwise unavailable. One major reason for the development of the bill was the interest in readjustment, a concerted emphasis on the reacclimation of veterans from a combat to a civilian way of life. Part of the concern was economical: the need for returning veterans to become part of the workforce again necessitated, the government believed, continued education and financial incentives for those coming home from the war.

The GI bill was a key contributor to the rise of the white middle class (though it also significantly aided members of other ethnic groups) and to the economical and educational establishment of postwar society. Educational opportunities were often technical, and American colleges and universities were turning out engineers, designers, and other technicians in vast and unparalleled numbers. The relationship between wartime progress and peacetime employment prospects was a close one. Technical industries had found a lucrative market in federal and civil sectors alike, and their role in the development of wartime technologies was frequently showcased. Take, for example, an advertisement in *Scientific American* for

Hevimet, a Carboloy metal alloy, which featured an artistic rendition of a battleship in the throes of combat at sea:

> They *stay* "on target" with Hevimet.
> Keeping an aerial camera or a naval gun "on target" is a tough proposition in a gusty sky or rolling sea. It is a job for accurate balancing and gyroscope controls . . . and therefore an ideal spot for Hevimet, super-heavy Carboloy created-metal. (*SA*, May 1952, 44)

This advertisement is a common example of how closely linked commerce, employment, and war were in the American mind. The freshness of wartime technology still lingered and was therefore an efficacious advertising technique, but it also reflected a patriotic pride in the United States' technological prowess—to the extent that product development and military application seemed a logical (and effective) marketing connection. This said, industry at all levels of the labor landscape had ballooned as a result of the war, and there was a certain implicit connection between war and prosperity that Americans naturally embraced. It would perhaps have seemed crude to explicitly suggest that war was good for business, especially in light of a half million American deaths during World War II, but this was an undeniable truth of the postwar years, and Americans were well aware of it. Things were good; things were good because war and capitalism are happy bedfellows. But perhaps this advertisement from the Atomic Power Division of Westinghouse Electric Corporation reflects the connection even more explicitly:

> How Would You Solve This Problem?
>
> You read a lot of magazines. You see a lot of ads. You've seen many of the printed messages from companies seeking trained scientific men. Everybody runs them. We do, too. And they help attract important and valuable men. But somehow they don't seem to measure up to the situation we have here. None of the usual words or phrases gives exactly the picture we'd like people to see.
>
> We Need Engineers
>
> You see, we have a contract with the Atomic Energy Commission. We aren't making bombs or turning out isotopes. We are building a nuclear en-

gine for a submarine—and our next big job is to build one for a large naval vessel. Maybe this sounds more like war work and less like putting atomic energy to useful work for mankind. But—the next steps will be atomic power equipment for peacetime purposes. (*SA*, March 1953, 36)

In fact, this does sound like war work, and it did attract "important and valuable men" to the industry—people who actively participated in establishing the companies that became major contributors to the economic expansion of the 1950s and the extremely audible sounding board for the Cold War ideology. The very language of the advertisement is a less-than-subtle indicator of how important getting the rhetoric right actually was: "None of the usual words or phrases gives exactly the picture we'd like people to see." The picture, it turns out, is one that carefully welds an inflexible joint between the benefits of military technological development and domestic economic stability. Imagine the excitement that a corporation like Westinghouse must have felt at securing a contract for such a project: not only would it make the company richer and more powerful than it already was, it would put it on the map as the business that developed the first nuclear submarine. Yet it was important also to craft the message in such a way that people (especially the "important and valuable men" who might actually have a civic conscience about such matters) did not see companies like Westinghouse as merely opportunistic war machines concerned more about the profit and profile such a contract would engender than the sweeping military involvement the contract explicitly set forth. "How would you solve this problem?" By suggesting that once the military incorporation of new technology makes the Free World impervious to outside threats, we can turn our attention to the "next steps," namely, "atomic power for peacetime purposes."

In a February 1953 *Scientific American* advertisement—this one geared less to the recruitment of scientific workers and more toward issues of domestic security—Fenwal Electric Temperature Control and Detection Devices appealed to the safety factor of technological development, asking the responsible question, "What do YOU want to protect?" Several items are listed as falling under the watchful eye of Fenwal's technology: "plane passengers and crew" (by providing reliable heater controls for temperature regulation); "lives, cargoes, and ships" (by supplying accurate fire alarm systems); "food at its flavorful best" (through the manufacture of thermostats in food-processing machinery); and "a modern metallurgical laboratory" (in which "Fenwal engineers are constantly developing and improving devices to help protect product, processes, property and people"—the alliteration being a nice rhetorical touch to help reassure the reader through a memorable list of items that are importantly secure) (43). In this example there are photos of happy airline passengers, a couple waving joyfully from the deck of a luxury liner, a mother and her daughter sweetly baking cookies together, and an engineer in a laboratory gazing competently at the dials and gauges of his profession.

The message, at least in terms of the advertising in popular science periodicals, was one of reassurance and opportunity, playing on the benefits of the superlative technology that marked U.S. superiority while concurrently underscoring the solubility ethos of the scientist/engineer and the safe, reliable, and beneficial aspect of state-of-the-art products, processes, and services. Some advertisements emphasized the direct and explicit connection between war and technology, while others only hinted that the military and the technological were reciprocal entities that needed to be exploited for fiscal stability and the peaceful furtherance of scientific innovations. Even more significant is the perception that war, technology, and the economy were so tightly interwoven that it was intellectually and ideologically impossible to tease them apart for rational and objective analysis. This is the basis for the entire Cold War mind-set: the terministic screen that dominated 1950s culture was one that allowed the language of prosperity and mechanistic advancement to seep through, but successfully distorted the dangers and pitfalls of the cultural/technological activity that defined American living. The dangers were sensed, but they were rhetorically deflected by the popularization mechanism that represented science as the competent intellectual body that guarded our most precious commodity: our way of life. Still, there was a persistent feeling of unease created by the sheer newness of the environment; the world was changing quickly, and this always threatens one's sense of stability and order, the very things science and technology professed to offer. Another of the ironies of this period, then, is that while science was philosophically based on the premise that universal laws are constant and unchanging, the products of the scientific method had created a world that encroached on people's innate need for security and constancy because it introduced changes that had a direct and nearly immediate impact on the way we conducted everyday life. Popular science adopted the conflicting role of both perpetuating anxiety about environmental problems and easing the tension through the cultivation of the solubility ethos.

Moreover, the affluence generated during the years shortly following World War II contributed to many strong and, at times, conflicting feelings that were difficult to reconcile. Domestic opulence, wartime economic jumpstarting, and technological progress intertwined to create living conditions that were both personally lucrative and socially unstable. American solicitude about this instability was the result of fears about global unrest, certainly, but it also had sources much closer to home. The sheer abundance enjoyed by the average American seemed disproportionate; it was difficult to imagine that so much was available to so many without some form of unknown and unforeseeable retribution. While there was no widespread ecological or environmental activism as part of the public consciousness in the way there is today, there was a vague sense that the United States was taking more than its fair share of the world's resources. Statistics to this effect were readily available to the readers of popular science journals. In the April 1953 issue of *Scientific American*, in the regular monthly feature "Science and the Citizen," the editors wrote the following:

> In the past four decades the U.S. has consumed more metals than the entire world had used from the beginning of the Bronze Age to the First World War. So says the Defense Production Administration in a recent report called Raw Materials Imports: Area of Growing Dependency. Although the U.S. still produces far more raw materials than any other country, it is now importing 9 per cent of its total requirements. The report lists shortages in 27 important raw materials, including:
>
> Asbestos: 95 per cent imported. Supply inadequate.
> Bauxite: 65 per cent imported. Supply adequate.
> Beryllium ore: 90 per cent imported. Supply very tight.
> Chromium ore: 99 per cent imported. Supply inadequate.
> Columbium (for high temperature alloys): 100 per cent imported.
> Supply scarcer than any other alloying metal.
> Copper: 35 per cent imported. Supply inadequate.

Nickel: 99 per cent imported. Supply critically short.
Tin: 100 per cent imported. Supply inadequate.
Zinc: 35 per cent imported. Supply adequate. (44)

The fact that as one of the largest producers of raw materials the
United States was still consuming more than it could produce should be
a compelling indicator of the rate at which Americans were sucking up
raw materials to keep pace with the lifestyle to which they had become
accustomed. But what rhetorical effect did the social unevenness created
by widespread wealth and the environmental/technological concerns
that accompanied it produce? How did science rise to the occasion, and
what expectations did the public have of science to address their most
pressing environmental difficulties? In the September 1952 issue of *Scientific American*, the editors printed passages from a five-volume report
called Resources for Freedom. The seriousness of the situation was put
this way: "'Consumption of almost all materials is expanding at compound rates and is thus pressing harder against resources which, whatever else they may be doing, are not similarly expanding,' says the
report. The [President's Materials Policy] Commission adds that the U.S.
must resign itself to importing an increasing share of its raw materials.
Their cost, it points out, is bound to rise. In the case of some materials the
shortage is already worldwide. The U.S. military services have laid out a
jet-plane program, for example, which calls for more cobalt than is available in the world" (70).

Aside from the obvious concern that the United States was demanding more than it could supply (in a rather grotesque display of U.S. opulence, the type of wastefulness and excess that made other, poorer
countries look at the United States with disgust and hatred, a phenomenon still prevalent today), the title of the report should be particularly
conspicuous. The connection between material abundance and freedom
is an important defining factor of American disposition toward the link
between capitalism, material acquisition, and the idea that these were
underlying principles in support of freedom. It should come as no surprise that in a country that idealized material consumption, physical
control, and capitalistic modes of conduct that science should be so central to the stability of these activities. In the absence of any but the most
superficial spiritual guideposts, science became the active ideological in-

stitution because it manifested exactly what we valued most. And in so doing, it aided in perpetrating material shortages that threatened our freedom because it threatened our ability to control the forces of nature and it created a vulnerability, an exploitable weakness in our super-power. This is made explicitly clear in the example about the jet-plane program: not only are shortages and overconsumption damaging our life-style comfort level, they are interfering with the development of military technology that has a direct repercussion on our safety and protection—in short, our freedom.

In another "Science and the Citizen" entry, this time in the July 1953 issue, editors introduce one of the first articulations of the now well-known greenhouse effect:

The Earth Is a Hothouse

By adding six billion tons of carbon dioxide to the earth's atmosphere each year, man's industrial activity is slowly warming up the earth. So says Gilbert N. Plass, physicist at The Johns Hopkins University. He told a recent meeting of the American Geological Society that he recalculated the opacity of carbon dioxide to long-wave heat radiation and found it to be much greater than had been thought. This means that carbon dioxide in the air acts like the glass in a greenhouse roof, allowing short wave heat to come in from the sun and preventing the escape of longer heat waves from the earth.

Plass estimates that at the present rate of industrial activity the amount of atmospheric carbon dioxide will double by the year 2080. That increase would raise the temperature of the earth by about 4 percent, and the gas will probably trigger a further rise by another process. By blocking the loss of heat from the tops of clouds and reducing the temperature differential between their tops and bottom, it will weaken the atmospheric convection currents that are responsible for rainfall. So the climate may become clearer and drier, allowing still more heat from the sun to reach the earth's surface. (46)

This brief entry may be one of the first times the global warming Plass describes was expressed in the greenhouse metaphor, an image that would change public perception of the earth's environmental state in subsequent decades. The metaphorical implications of scientific rhetoric will be handled in the next chapter. Here, it is important to note that the scientific issues presented to the public were not limited to nuclear war and atomic energy, but were, in fact, beginning to touch American life and thought on many diverse levels.

As mentioned elsewhere, the ambivalence experienced by Americans was the result of the contradictory nature of scientific activity. On the one hand, the very industry that was responsible for the current economic upswing was also the cause of some very real and very serious environmental problems. Efficiency in industry meant that plants were larger, louder, dirtier, and more prevalent than ever, and this efficiency could be attributed to the improvements in excavating raw materials, producing usable materials, manufacturing products, and shipping goods that modern technology made possible. The trade-offs were pollution, suburban sprawl, overpopulation, crowded roads, energy crises, and resource shortages, to name only a few. People were beginning to notice that their recent economic well-being was not without cost. And science was both the problem and the solution.

Finding solutions for everything from new forms and applications of energy to innovations in medicine required research. One of the most common practices for product and medical research was the use of live specimens to test new drugs, to determine the safety of household products, and to research diseases. Experimentation with animals was a well-established scientific practice, one which had its detractors even in 1953. However, what political activity existed against animal experimentation was viewed as the reactionary mumbo jumbo of a few antiprogressive kooks. In the following "Science and the Citizen" excerpt, this also from the July 1953 *Scientific American*, the editors clearly revel in the assumption that the antivivisectionist cause was all but disappearing:

Antivivisectionists in Retreat

The antivivisectionist movement in the U.S. has been reduced from a national campaign to mere local harassing actions, says the National Society for Medical Research. The society reported a major change in climate, with the antivivisectionists instead of scientists now on the defensive.

The Society, which has campaigned for the cause of animal experimentation since 1946, says that the Hearst newspapers have given up the antivivisectionists' cause. Its libel suits on behalf of scientists against the Chicago Herald-American were settled out of court early this year. The publisher announced that the paper's editorial policy had changed with the death of William Randolph Hearst. Since then, says the Society, the paper "has been giving . . . animal experimentation . . . a straight play." Another indication of the change in policy was the New York Journal-American's virtual silence during the 1952 debate over the Hatch-Metcalf Act, which made animals available for experimentation.

Antivivisectionists are concentrating on making animals difficult for scientists to obtain. The American Humane Association, the local clearing house for local animal welfare groups, is campaigning to defeat legislation which would provide animals for research. But most medical schools are now able to get animals under state or local laws or by working arrangements with humane society pounds. The only places where difficulties remain are Boston, the District of Columbia and Pennsylvania. (48)

From our more "enlightened" vantage point—in our age of self-congratulatory tolerance, civility, and humaneness—the idea that some groups might actively object to experimentation on animals seems reasonable and desirable, even imperative. The notion of animal rights is a truism today: animals are innocent in the sense that they are not burdened with moral concerns about their own behavior. They simply do what is natural for their species to do. Just as human beings have natural, uncompromising access to rights of person and property, so too should nature's creatures enjoy an unmolested, unencumbered existence. The contemporary hierarchy suggests that animals share a place with humans in terms of natural rights; they are, in fact, more noble because they are pure, uncontaminated from the stain of civilization. Rabbits, guinea pigs, rats, hamsters, dogs, cats, nonhuman primates such as monkeys and chimpanzees, and other domesticated animals all have their place in the collective ethic, and torturing and killing them in the name of science seems to many people unnecessary and cruel, especially when there appear to be alternatives (albeit often expensive and time-consuming ones) to using live specimens for vivisection. There are, of course, many who still feel animal experimentation is a necessary evil, that human progress (especially in the name of disease eradication) takes priority over animal rights. But even here, many are reluctant to accept animal experimentation when alternatives are available. The ethic has shifted in recent years, in large part due to the efforts of politically active environmental groups such as

Greenpeace and the Antivivisectionist League, to ease the encroachment of human activity on the biological realm. Animals enjoy an enhanced status today, one which helps protect them against the anthropocentric will of humankind.

But such an ethic was in a quiet infancy at the time, and the attitudes regarding animals paralleled the attitudes humans (especially Americans) had about their place as the governors of nature. Science is by its quintessence a domineering epistemology; the very principles under which it operates demand a certain willful manipulation of the natural world. To understand it, the scientist must dismantle it. To dismantle it sometimes means to destroy it. This was viewed as an obvious—if unfortunate—residual effect of scientific activity, and it was done only so that, by better understanding our natural world, our own species could adapt and flourish as our destiny dictated. The editorial remarks about antivivisectionists transparently evidence this hierarchical chain of being. Clearly, science had a right that superseded any dumb brute's, and antivivisectionists, with all their moral posturing and systematic irritation, were only standing in the way of a greater good: scientific progress. Where scientific ethics were concerned, animals did not fare well; while scientists had given a great deal of thought to the implications of macroscopic natural transformation—such as a nuclear holocaust and the disintegration of most life on earth—the subtler points of ethical accountability were still largely ill-considered. It seemed obvious that in order to help humankind science would have to sacrifice other life forms. It was simply a matter of priority, and from a rationalistic point of view, the choice was certain. By banishing the antivivisectionists to the periphery, the editors of this "Science and the Citizen" entry dismissed them as antiscientific and antiprogressive, thus marginalizing such efforts as silly irritations that would soon heal.

The language bears this out. The antivivisectionists were in retreat—that is, on the run—outlaws staging a coup that had been successfully thwarted. They had been reduced to "mere local harassing actions." There were only a few cities where difficulties remained. This was a mop-up operation, with scientists as the governing force of the victorious army. Once more, pervasive military language represented scientific activity, as if the concepts of war and technological progress were inseparable, at least insofar as such activity was showcased by scientific popularizers for public consumption. The solubility ethos was further reinforced through a demonstration of military competency, a characteristic evidently highly prized among Americans. U.S. citizens had voted in a celebrity general as president in 1952; they had benefited greatly from the spoils of wartime victory; they had seen a deep connection between science and military technological application; and they had equated national strength with the ability to wage war successfully with state-of-the-art weaponry. Americans are a competitive lot, and war is the ultimate game. As in any professional sport, it was necessary to have the latest equipment in order to be in full form. Science supplied the equipment, but it also reminded us that the pragmatic implementation of such equipment had its use in war. What was not used directly for war almost always sifted its way down to the public through military channels—often technology that was deemed of no use to the military became marketable for domestic use. Silly Putty was a failed experiment for a high-temperature gasket in jet aircraft; plastics of many varieties had similar fates. Slinkies were originally designed as springs in military mechanics, and microwaves were used in radar experiments until scientists discovered that the heat they generated would make functional ovens (the first microwave oven was the Amana Radar Range). Technological, military, and domestic growth were one entity.[2]

Antivivisectionists, then, were merely one nettlesome obstacle to be overcome to reach a desired objective. The political dimension of environmental activity was closely linked to the martial orientation science naturally adopted as a result of its close ties with the military. In other words, the tensions groups like the antivivisectionists and the scientists experienced were approached by both groups as an ideological conflict rather than a reciprocal discursive interaction. Courtrooms, in this case, were the battlefield on which the conflicts were resolved, but there was little discursive compromise: either animals would be used for experimentation or they would not. This example reflects an interesting and dominant orientation, since the one hegemonic perspective these groups had in common was the idea they were engaged in a warlike skirmish, the outcome of which would establish one ideology over the other. Such a vantage point parallels the country's overall mind-set regarding the larger ideological conflict that was the Cold War. The antivivisectionist cause was a battle over ideas, albeit ideas that had a practical outcome, but the antivivisectionists represent an ethical impetus that was beginning to gain minor but discernible momentum. While the battle may have been small from the corporate/governmental/scientific point of view, it was clear that scientific practices were being questioned by certain peripheral segments of society.

Successful reform generally takes place in increments. It is rare that a full-scale revolution can take hold, even when the climate is favorable for it. Most of American society still viewed science with enough awe and reverence to grant the scientific body authoritative action in matters that affected the country. The antivivisectionists failed during this time for the simple reason that they were viewed as a rogue minority who had radical ideas about the inalienable rights of animals, and they were up against an entity that most felt had far more knowledge and wisdom about the broader and more informed notion of progress than some marginal faction did. As a matter of national priority, their cause was doomed since science had a foothold on the American way of thinking, which was not going to be easy to pry loose, and most people had no stock in doing so anyway. The country was into the Cold War crisis too far to relinquish science's expertise in situations that only scientists could understand and problems only scientists could solve. If a few rats, rabbits, dogs, or monkeys had to die in the interest of scientific exploration, then no over-the-top leftist group was going to prevent that from happening.

These years were a conservative period in American history. Conservatism seems to thrive during periods of economic prosperity and stability because the public has an investment—literally and figuratively—in keeping things the way they are. Change is suspect in such an environment, and Americans were not receptive to any more change than they had already experienced during the postwar years. Dwight Eisenhower himself was the embodiment of the conservative parental figure—the father-knows-best image that pervaded American moral, ethical, and political thinking. The patriarchal reassurance that Eisenhower personified virtually guaranteed the widespread acceptance of the institutions he valued himself, and this certainly included the scientific and technological community. At the risk of begging the question, we might observe that Eisenhower was in many ways the quintessential Cold War leader. His military background was just about as impressive as a career army man could expect, and unlike like his Pacific campaign counterpart Douglas

MacArthur, he was a shrewd, resourceful, reserved, even wise, politician. His well-known assertion in the so-called Chance for Peace speech that "every gun that is made, every warship launched, every rocket fired signifies, in the final sense, a theft from those who hunger and are not fed, those who are cold and are not clothed" demonstrates this oddly soothing, fatherly characteristic in a rather curious way: by suggesting that military armament is, in fact, robbing the world's people of basic necessities, he is articulating the irony and ambivalence of the Cold War situation in one terse, uncompromising statement. More than that, he is suggesting that the world's people are in his (or, at any rate, America's) charge, and that it is our responsibility to ensure such thievery is not tolerated. Yet, in another claim from the same speech, Ike says the following:

> This world in arms is not spending money alone. It is spending the sweat of its laborers, the genius of its scientists, the hopes of its children. The cost of one modern heavy bomber is this: a modern brick school in more than 30 cities. It is two electric power plants, each serving a town of 60,000 population. It is two fine, fully equipped hospitals. It is some fifty miles of concrete pavement. We pay for a single fighter plane with a half million bushels of wheat. We pay for a single destroyer with new homes that could have housed more than 8,000 people. This is, I repeat, the best way of life to be found on the road the world has been taking.

Where does this paradox leave us? Where do we stand as a nation when the reality of the global situation demands this cost? From a practical point of view, the conservatism that characterized the early Cold War years was an attempt to come to grips with these glaring inconsistencies, to impose some perception of order on a world ripping apart the seams of our most cherished ideals and preconceptions. Now was not the time to question our approach or our leaders, because now, more than any time in nearly a century, our nation faced a daunting crisis. Rogue elements were not welcome because they signaled uncertainty and fluctuation at a time when people craved stability—even if that stability was a rhetorical construct and a political illusion. Political radicals like the antivivisectionists were merely irritating an open wound; they were political pests shouting slogans for a few mice when the biological well-being of the entire planet was at stake. It is small wonder that they were dismissed as petty annoyances. They seemed to lack perspective, and perspective could only be supplied by those who knew what they were doing (read: those who helped arrange the situation in the first place). Eisenhower was as committed to science as anyone because he assumed that science was the only hope for a world that had brought itself to the edge of an abyss on the shoulders of misdeveloped and misapplied technology. He recognized the cost of using scientists to create new, more destructive weapons, but he also realized that the technological strides being made would have no equal during a more benign period.

The situation was so remarkably complex in part because of the ideological substructure that supported it. It was clear that the clash between the two paths Eisenhower described in his speech necessitated sacrifices and even unthinkable risks, and his rhetorical role was equally clear: in an era of looming crisis, he had to carry the burden of strength and guidance for an entire nation, and the American public looked to Ike as a friend, a mentor, and a father. His station in the American macrocosmic nuclear family helped make sense of the ambiguities that threatened the ideals that carried the United States to victory in the greatest war the world had ever seen. But the victory, tenuous and fleeting as it was because of an ambitious, overzealous, and jealous empire (so it was portrayed), was uncertain and somehow incomplete. Eisenhower valued progress and trusted science to help guide us through the murky waters of sociopolitical discord between superpowers, but part of the problem was that there was no point of reference, no historical precedent off of which to sound these new experiences. It was a historical frontier that the United States (and the Soviet Union) were traveling through, as frightening and as dangerous as—and perhaps more consequential than—a trip to Mars. The only group equipped to face the challenge with courage and knowledge were the scientists; the only administrative body with enough clout to back their odyssey (and the public's) was a militarily staffed White House. Like children who are dubious about the prospect of diving into the deep end while their parents urge them on, U.S. citizens now had to simultaneously trust and suspect their intellectual and political leaders.

Public perceptions of and reactions to Cold War events were as ambiguous and varied as the events themselves. Public response was not so uninformed that political and scientific activity was simply and completely given over to those who moved about in the grim drama of international affairs, but practical measures dictated by the authorities were not actively and consistently challenged to promote change either. Anxiety prompted questions from the American public, and they almost always concerned prospects for the future: how do we handle the population explosion? How do we feed the masses? Where do we find new forms of energy or tap into the resources needed to supply the demand? What are the natural consequences of our increased need? People began to question the extent of their impact on the earth, knowing instinctively if not empirically that the natural resources of a finite planet are indeed limited, and that they were exacting a perilous toll upon the very materials needed to maintain an opulent way of life.

As people became increasingly concerned over the unprecedented population explosion, they also fretted over potential food shortage problems that loomed on the horizon. Never before had the United States so many people to feed, house, clothe, and legislate, but the lack of food took priority over all these related issues. Projected food shortages were a special preoccupation with people, and new and innovative ways to feed the masses were frequently proposed in popular science magazines. In 1960 Arthur C. Clarke wrote a brief article for *Popular Science* called "Will a Hungry World Raise Whales for Food?" Taken from Clarke's book *The Challenge of the Sea,* this article taps into a special source of anxiety for Americans: overpopulation and food shortage. What makes the proposal especially quaint from an enlightened twenty-first-century perspective is that such an idea would be politically unthinkable today, especially since Clarke proposes not only hunting whales—he acknowledges that "some 50,000 of these great animals are killed every year, providing valuable oil, meat, and other food-products" and that "you may never have eaten whale meat, but when properly prepared, it tastes very much like beef"—but also raising them as cattle: "For whales are cattle, even though they weigh a hundred tons or more. This has long been recognized in the use of the term bull, cow, and calf in connection with them. They are intelligent creatures who can communicate with each other by underwater sounds, so it should not be difficult to direct or control them, perhaps with additional help from electric fields" (75-76).

Clarke proposes this idea, "if only because more familiar kinds of meat will be getting steadily scarcer and more expensive" (76). A well-known scientist and writer, Clarke is perhaps the archetypal example of the popular scientist. While he had done extensive work himself on radar and other forms of communication technology, he was also an avid writer and advocate of scientific popularization. His celebrity lent him an authority that lesser-known popular scientists lacked, and this meant that the general public viewed his scientific assessment and his proposals with interest and approval. His rhetorical recognition that language dictates reality—that whales are cattle because we call them cows and calves—is interesting because it implies that biological classification (a linguistic activity) is the bottom line in the legitimization process. By today's standards, the whale proposal seems inhumane, even uncivilized, but that is mainly because we know more about the true intelligence of whales and find treating them with the same disregard we show toward other livestock unthinkable. This aside, clearly the proposal never came to fruition, yet it exhibits the kind of problem-solving freedom given to scientists to think aloud in this manner as a way of addressing some of society's more pressing issues.

The argument he makes is almost dismissively practical; he refers to the whales as "floating gold mines." A single whale, he says, can bring as much as $30,000 on the open market, and taming whales would be easy because of "our modern resources" (96), by which he apparently means our technological capabilities as well as the scientists needed to steer them. One striking attribute of this article, and most *Popular Science* articles, is its amazing superficiality. It runs a mere four pages, and even these are liberally supplemented with illustrations and photographs. *Popular Science* is to community science publications what *USA Today* is to serious journalism. The visual nature of the piece was undoubtedly deemed necessary for a broad (and intellectually limited) audience, and the depth of Clarke's proposal hardly merits serious consideration. Yet one of the reasons for Clarke's popularity was (and still is) his ability to discuss scientific ideas in layman's terms—to avoid alienating people through the use of overly technical language and to treat them, in some shallow way, as equals. One wonders just how seriously Clarke himself took this idea; given the sheer frivolity of the proposal—one almost feels that he is making this up as he goes along—it appears it was just a quick way to make a few bucks. But it is not likely that it was viewed this way by the public, and herein lies its importance as a historical and rhetorical document: by creating a dialogue with a general readership on issues of a scientific nature, *Popular Science* could encourage people to feel involved in the decisions and concepts scientists were wrestling with. The problems are global, or at least national, if sometimes a bit overblown, but by letting the populace into the inner sanctum of scientific solubility, people could feel as if they were in the loop, aware, and empowered. The public relations effect these publications and their stories had was great even if the true level of public participation was minuscule. It created the sense that steps were being taken by the people best equipped for the job, and even if the participation was illusory, it created a calming effect on a nervous laity.

Still another growing problem that might seem unlikely in this context of prosperity and technological advancement—but one that reveals the true scope of people's ambivalence and need for escape—was the rising rate of drug usage, especially narcotics and amphetamines. A full-length feature in the October 1959 issue of *Popular Science* provides us with "The Tragic Truth about Taking Dope." The front page of the article shows a photograph of an addict sitting on the edge of a hospital bed with his head in his hands, wrapped pathetically in a thin wool blanket. Underneath, the caption reads: "In a New York hospital, a narcotic addict goes through misery as he tries to 'kick the habit'" (81). As unlikely as it seems, this is yet another scientific problem demanding a scientific solution. Following the brief introduction, in which an addict's life is reduced to "misery," the author, Herbert Yahraes, cannot help but editorialize by observing that the "dope addict brings family shame and anguish. He finds it increasingly difficult to get any real joy out of life. He no longer even gets the temporary kick from dope-taking. He is likely to wind up among underworld characters—or in prison. He suffers, really suffers, physically" (81). This becomes a scientific problem for two reasons: one, the development, manufacture, and availability of narcotics is more efficient in large part because of advances in chemistry, medical technology, and cheap processing and delivery methods. More significantly, however, drug abuse is viewed as a medical, psychological, and social problem, meaning that doctors, psychiatrists, and other scientifically trained technicians and specialists are the primary authorities not only for diagnosis, but also for treatment and reentry into the civilized world. Doctors have had a long history of contact with the mentally ill, but

their role was one mainly of monitoring and confinement of the "luna-
tic," whereas here there is an emphasis on the physiological and emotional cause of the addiction. *Popular Science*, as one might expect,
focuses on the scientific processes used to understand the addict better
through studies, data, and experimentation: "Experiments at the drug-
addiction hospital in Lexington, Ky., have shown that even persons 'on
the nod' will quickly rouse themselves when asked to do something and
will do it with customary skill. They may do it a little less speedily than
usual, and on psychological tests they score a little lower than usual. But,
states a report to the American Journal of Medicine: 'If sufficient supply
of the drug is available, the overt behavior of the addict is not unusual
and he can carry on a highly-skilled technical occupation in a fairly satisfactory manner'" (82-83).

There are two points of interest here. First, the methods for determining the significance of addictive behavior are scientifically oriented,
underscoring the importance of clinical testing for defining and charac-
terizing the problem. Second, and more subtle, is the preoccupation with
the addict's ability to do work, especially work in a "highly-skilled technical occupation," which exposes a connection to the stress placed upon
the need for a scientifically trained, reliable, competent, and technically
proficient workforce. It is no coincidence that the testing of the addict's
work habits centers around his (or her) ability to contribute to the overall
well-being of the country, and this translates into a technical acumen regarding the advancement of science and technology. Hence the drug
problem is a social problem, and one that touches science on a number of
levels. Not only is it an issue to be defined and monitored by science, one
that science helped to create; it is also a problem that affects the future of
science as a centralized player in international politics. Without a solid
and self-sustaining pool of qualified technical workers, the United States
would surely fall behind the Soviet Union in technical fields paramount
to the stabilization of the country and the protection of the nation (and
others) against communism. The article goes on to discuss what one
might expect: the degeneration of the addict into a financially and emo-
tionally crippled criminal, seeking a fix wherever and however he can—
robbing, stealing, even killing in his mindless pursuit of drugs. The gravity
that the drug problem is given is appropriate, but the underlying message
marks the impact this question has on the fate of the nation. The perva-

sive rallying cry that denoted the early Cold War demanded that everyone contribute to the collective strength of the nation to stave off the crisis. People on drugs, such messages suggest, not only hurt themselves and their family, but also society. They are unpatriotic.

The battle against alcoholism took on a similar perspective. In the September 1952 issue of *Scientific American*, it was reported that "in 1948, there were almost four million alcoholics in the U.S., about one million more than in 1940, according to estimates made recently by E. M. Jellinek and Mark Keller of the Yale Center of Alcohol studies" (40). The editors added that "reported cases increased in every section of the country," and that "for the past 10 years Nevada, California, and New York, in that order, have been the three most heavily affected" (40). New York was a primary industrial state, producing and transporting a great deal of the country's goods, and New York City was a central financial trade center, a key location for international and diplomatic relations. It also was the chief city in the buying and selling of technological stock. California developed and manufactured a significant portion of the technology commonly used by both the military and the American public. And Nevada, like much of the Southwest, was a locale for nuclear testing and experimentation. Is it a coincidence that such states should see an increase in alcohol usage? Did the anxiety created by increased workloads, hours on the job, and a frantic scrambling to stay ahead of the Russians technologically and commercially take its toll in the form of alcohol abuse? Did the need to escape the realities of the postwar world parallel the intake of spirits, giving the American workforce the numbing effect it needed?

The apparent rise in alcoholism can be accounted for in a couple of ways: it is possible that the interest in finding and diagnosing alcoholics increased during this period, thereby suggesting that alcoholism did not rise at all, but the other possibility is equally compelling. Like the increase in drug use, a rise in alcoholism could be attributed to people's need to ease the distress of the nuclear age. The uncertainty principle can have an exacting effect on the human consciousness, especially if one's cultural upbringing gives weight to the pursuit of order and certainty. During the war, the goal was clear and unwavering: beat the oppressive Axis powers into submission. People were willing to work hard and make sacrifices in order to accomplish this unambiguous task. During the war, there was no intangible, abstract, sociopolitical theory to strive toward, save for the ubiquitous (and rather vague) notion that they were fighting for freedom and against tyranny. The Cold War was different, however. People harbored a sense that democracy was a superior form of government to communism, but if challenged, few people would be able to provide a satisfactory account of the principles that guided these ideas. Communism was simply evil because it was oppressive and it interfered with liberty. Democracy was better because it gave everyone a vote and was ostensibly laissez-faire. But because there were no outright hostilities, there was an underlying sense of pointlessness and confusion about the real purpose behind the indefinite exertion that seemed to be infecting the nation. For a pragmatic people, this is a recipe for neuroticism. Without the structure, order, and purpose that defined the war effort, people operated with equivocalness and foreboding. And for those impious few who had difficulty comprehending and openly embracing the scientific rationale driving the Cold War effort, alcohol must have seemed an effective alternative.

This was also the era during which alcoholism began to be widely viewed and described in terms of a disease rather than a weakness in character or a straightforward act of self-destruction. In the years following the repeal of Prohibition, the medical industry's conceptualization of the alcohol problem changed drastically; the focus was no longer on the substance itself, as it had been in the past, but rather on the affliction that was the bane of the alcoholic. During earlier temperance movements, the advocates of Prohibition and abstinence targeted alcohol itself as the primary menace, seeing the substance as evil and thereby creating a relatively easy target to eradicate. While it was difficult to rid the world of drunks—being people who had wills and rights—it was a much more manageable task to rid the world of the source of the problem: alcohol itself. However, as Prohibition had shown, this was an ill-conceived strategy. Alcohol usage during Prohibition not only increased but also created a thriving underworld of crime and bootlegging that was impossible to regulate and laws that were impossible to enforce. More than a few petty street thugs made their fortunes through the illegal manufacture, procurement, transport, and distribution of alcohol. Alcohol itself, it seemed, was not as easy to erase as the temperance advocates had imagined.

With the repeal of Prohibition it became necessary to redefine the problem; hence the emergence of alcoholism, giving the problem the status of disease and focusing efforts to solve the problem on the alcoholic himself. Aside from the necessity of this move—it was clear that the direct approach of Prohibition had failed—it had other advantages as well. As a disease, alcoholism required greatly expanded treatment and research efforts to create a contrast to what Selden Bacon referred to as the "Classic American Temperance Movement," and it produced an entire industry of health professionals and medical researchers to study the problem. According to Ron Roizen: "The modern alcoholism movement split cultural 'ownership' of the alcohol problem domain between, on the one hand, Alcoholics Anonymous (AA), a voluntary fellowship devoted to the rescue and spiritual renewal of fellow alcoholics, and, on the other, a mainstream scientific enterprise devoted to promoting the importance of research in addressing the nation's alcohol problems" (2).

It is not my intent here to debate the politics, efficacy, ethics, or even "reality" of the shift in the alcohol paradigm from alcohol as an evil substance to alcoholism as a chronic but treatable disease. It is interesting, however, that where the antiscientific approach of alcohol eradication had failed, medical science stepped in to embrace the problem. Alcoholism was a disease, or even an allergy, like any other chronic malady, thereby discharging in one notable move years of supposition that the alcoholic was a weak-willed, hopeless loser and that the only way to cure the alcoholic was to remove the alcohol. As Roizen puts it: "The disease idea also offered destigmatization to the alcoholic and a measure of new symbolic legitimacy for beverage alcohol itself—which, in the new paradigm's lens, harbored little more responsibility for alcoholism or alcohol-related troubles than did sugar for the disease of diabetes" (3). While such a vantage point does offer a level of humaneness in the clinical and personal treatment of the alcoholic, it does not appear to have been an entirely altruistic move on the part of medical science. Rather, it helped provide yet another foothold for the scientific community to not only increase its authoritative status through the solubility ethos, but also secure funds and create jobs for research as well. As the inheritors of the disease concept of alcohol and drug abuse, treatment of alcoholics and addicts is big business, and it is a business with an investment in the scientific remedy. A sick society requires doctors, nurses, emergency medical specialists, lab technicians, pharmacists, drug manufacturers, psychologists, and medical therapists of all sorts to cure the disease. Roizen continues:

> Not the least remarkable feature of the modern alcoholism movement was that it in effect represented a trial run for the proposition that modern science could and should take charge of a major American social problem. If Prohibition had been a "noble experiment" in grand-scale, legislatively-imposed social uplift, then the modern alcoholism movement represented a bold test of a new, would-be, post-Repeal scientific hegemony. And although it was not science at all, but AA's spiritually oriented approach, that provided the movement's all-important evidence that alcoholics could in fact be helped, the disease concept's message of medicoscientific naturalism defined and premised the new cultural sensibility and spawned considerable research and treatment enterprises that would emerge over the remainder of the 20th century. (3)

While Roizen separates the alcohol treatment camps into the scientistic on the one hand and the Alcoholics Anonymous spiritual approach on the other, there is an even more interesting connection in the convergence of spirituality with the new alcoholism paradigm in the form of AA—an effective marriage of science and religion that both helped the alcoholic view the problem as one beyond his or her control and provided a spiritual outlet for his or her tortured soul. The alcoholic needed a surrogate, something he or she could turn to besides the bottle for comfort and strength. Hence the spiritual dimension of AA and the giving over of one's life to his or her "higher power." But this was only possible given the new disease orientation, since the alcoholic must feel unburdened of the responsibility for his or her malady in order to fully capitulate to an entity outside personal control. The very first of the famous twelve steps, in fact, asks the alcoholic to admit he is powerless against alcohol. Such an admission would not have been possible under the old, demon rum model since alcohol itself was seen as an irresistible force as long as it was available. Alcoholics were simply drunks who could not control themselves against the pull of booze: remove the source of weakness and one removes the problem. The disease orientation, however, provided the alcoholic with hope in a world where alcohol was not going to go away, and the best way to accomplish the goal of permanent sobriety was for the alcoholic to see the problem as a disease that only diligent personal therapy and a surrender to God could overcome.

The alcohol problem—and the alcohol solution—required yet another instance of scientific intervention. Like the drug problem, alcohol was a menace from a national perspective, not so much because of the lives it wrecked, the relationships it destroyed, or the health dangers it posed, but because it interfered with a national agenda that required everyone to be sharp, sober, driven, focused, and generally unencumbered by the haze of inebriation. Once again, science defined the problem in terms it could deal with so that it could get a handle on the problem and be the central agent for dispensing solutions. Studies conducted by the newly created Research Council on Problems of Alcohol fashioned the issue in terms of medical research that, at least early on, did not hold with the hypothesis that alcoholism was in fact a disease, yet despite this the paradigm endured. Alcoholism was deemed a problem in the first place largely because it threatened the well-being of the state and its single-minded pursuit of scientific and technological productivity. Even today, drug-, alcohol-, and tobacco-related health statistics are passed on to the public in terms of "work hours lost" or "insurance money spent," as if these were the bottom-line measures of the problem.

Such is the obvious orientation of a materialistic, pragmatic democracy: with or without the Cold War threat, Americans tend to view the impact of any given issue in terms of its effect on acquisition, labor, economics, and production. This is a phenomenon that John Dewey referred to as occupational psychosis. The interests of a social unit—on the most basic level, survival interests, modes of food production, and the unit's relationship to its natural environment—dictate not only the preoccupation with the necessities of living, but also the symbolistic, religious, ceremonial, and humanizing manifestations of that social unit's culture. For example, if a tribe lives near the only river within 100 miles and relies on that river to grow wheat and raise livestock, it is natural for a symbol-using culture to revere water, grain, and livestock and for these things to be abstracted to the point where they not only carry their literal meaning but also figure prominently in all aspects of that social unit's symbolic behavior. Such items may be important for marriage ceremonies or for rites of passage. The tribe may worship water gods and grain goddesses and view cattle not only as a source of food but also as a key element in the life cycle and perpetuation of the tribal unit.

In a more complex civilization—one that is based not only on networked material production and trade but also on governmental principles that dictate ideologies and civil conduct—the abstractions become more complex. Occupational here means not only the vocational activities that dominate a society but also the degree to which the material prerequisites occupy the mental, rational, emotional, and symbolic space of social life. Psychosis in this context does not mean mentally ill but refers rather to the mental state that prevails, what Burke calls a "pronounced character of the mind" (*Permanence and Change* 40). In the postwar American society, the occupational psychosis of capitalism dominated (as it still does today). When this economic ideology is coupled with its reliance on science and technology to thrive, the result is a pervasive insistence that all areas of social, cultural, fiscal, and national activity is contingent upon this way of thinking. Nearly everything is defined in terms of nationalistic fidelity and economic stability, which, as a society reliant on technological progress to thrive, only science can control. Alcoholics and drug addicts needed help for humanitarian reasons, certainly, but the real interest was in maintaining a strong, vibrant, productive technological workforce, one that would help ensure our ideals and preserve our way of life. In short, the American occupational psychosis of the 1950s (one that is still very much with us today) demanded that science take on the problem because the future of our scientific community—and therefore our national identity—hung in the balance.

The Symbolic Megaphone of *Life*

The very title of *Life* magazine suggests that it is a manifestation of the energy and texture American society comprises, a definitive document of the most important and compelling events in the nation. As a photojournalistic representation of American activity, *Life* is pictorial in nature, dedicating more space to images, photographs, and impressions than to words. While there are feature-length articles, they are sparse, and the reader quickly notices that most of the words function only to describe and underscore the images. The magazine clearly operates under the photojournalist's premise that a picture is worth a thousand words. More important, perhaps, is that the effect of this format is an imagistic symbolism, a sense that the vibrancy of the American landscape, buzzing with domestic activity, is captured in a few laconic, comprehensive photos. It announces the American way of life boldly and unambiguously, a compact representation of American ideology writ large.

For this reason, the American public looked to this publication for an overview of the state of the nation. Its short features and larger-than-life photographs made it the conclusive source of information that tightened the outlook and values Americans held dear. Just as large and revealing were the advertisements that, even by today's standards, dominated the pages of *Life*. There are few advertisements that take up less than an entire eight-by-twelve-inch page. Some advertisements spanned two adjacent pages, making for a folio effect difficult to ignore. Little of the advertisements' contents is actual text; instead they are littered with colorful, eye-catching illustrations enhanced with only enough words to carry a rhetorical message. Magazine advertisements of the 1950s have a signature all their own, unmistakable for the heavy reliance on illustrations rather than photos, the script font often used to give that "ringing endorsement" feel, and the tendency toward obvious hyperbole. Absent are today's computer enhancements and air-brushed photographs, and when viewing the advertisements, one gets an unmistakable sense of nostalgia, a gut-level feeling that these times were somehow more innocent and naive than our own. This is an illusion, of course. The America advertisers wished to portray was white, suburban, well-adjusted, well-groomed, and conformist. It is rare to find ethnic variety in the advertisements, unless the product wanted to further a stereotype of some sort (pictures of Chinese laundry workers to endorse a detergent, for example). These advertisements project the illusion that America is just as uniform, just as safe, just as homogenous as the people who populate the advertisements themselves, and it is yet another—perhaps deliberate, perhaps unwitting—effort to show the Cold War public that their anxieties were unwarranted, that everything was as chipper and carefree as the advertisements suggested.

Advertisements using science as an authority to sell products began to show prominently during this time, a rhetorical device of advertisers still with us today. Not all products fare well using science as a means of persuasion, but many unlikely examples persisted anyway. Scientific endorsement was used to sell everything from hams to toothpaste, and this vehicle fitted in nicely with the American predisposition to value and respect scientific authorities. Processed foods, medicine, automobiles, home improvement and domestic maintenance products, power tools, lawn and garden care, and nearly any other product one can name that would benefit the suburban homeowner were showcased using science as the ultimate consumer official. If science said the product would perform as promised, the public was likely to believe it. The scientific community, of course, had little direct interest in selling Bromo-Seltzer, but it certainly was in the best interest of the scientist to have the backing of a successful corporate bankroll. While scientists themselves were not seen gracing the pages of *Life* advertisements (Robert Oppenheimer or Albert Einstein did not, to my knowledge, ever sell Motorolas), science was frequently invoked as the final word on a product's worth, and this helped perpetuate the solubility ethos and ultimately created conditions under which scientists could have corporate jobs that might not otherwise be available.

The Bromo-Seltzer advertisements are worth a look, if only to provide an example of a trend that will persist throughout the course of mid- to late-twentieth-century advertising. In the December 15, 1952, issue of *Life,* a full-page advertisement features the face of a nurse peering reassuringly over a glass of bubbling Bromo-Seltzer. The copy above her reads, "Bromo-Seltzer Best for 9 out of 10 Headaches"—a simplistic appeal to the power of statistics, which influences Americans more than almost any other form of evidence. Random numbers like these are indeed persuasive because they create the illusion that a finding has been carefully and scientifically determined. Americans trust doctors and medical professionals because they have been trained to study the effects of medicines and treatments. Readers almost forget that they are reading an advertisement; they are drawn into the clinic with the doctors and see the results before their very eyes. This is, of course, a matter of rhetorically decorating the advertisement with people who possess the solubility ethos and who are, in all likelihood, models posing as medical professionals. One photo shows a doctor scribbling meaningfully on what appears to be a chart while another doctor and nurse examine a patient. This re-creation of what seems to be a familiar and plausible scenario adds to the sense that the findings were by no means arbitrary,

and that the effects of Bromo-Seltzer have been conscientiously and pro-
fessionally documented. Advertisements are perhaps the best evidence
that a certain mind-set has bled over into popular culture, if only because advertisers are very shrewd at discerning social trends and ex-
ploiting them for persuasive effect. It is clear that this approach works
because it reassures the American public by showcasing the experts in
their element. Symbolically, it functions as a surrogate for an actual visit
to the doctor; readers can self-medicate confidently because doctors and
nurses have already done the homework for them. Readers recognize the
familiar nurse's cap and uniform and the doctor in the long white coat.
Doctors and nurses are shown performing the expected rituals, which
are captured in several carefully staged images. The public's faith in the
medical profession is manipulated to sell a product, and while medicine
should certainly fall under the supervision of medical professionals,
there is no unequivocal evidence showing that these are professionals or
that the findings are accurate and complete. Nonetheless, consumers are
comforted because the visual impulse of the advertisement mollifies
them by lending credence to the product and convincing them that its
use is warranted and "certified."

The advertisement continues: "Doctors and Nurses Know How
'Nerves' Can Cause Headaches" and "Clinics prove that Bromo-Seltzer is
effective. Recent clinical tests show that 9 out of 10 headaches are associated with nerves—as most headache sufferers have known all along. In
published literature, headache specialists and neurologists substantiate
this medical fact." There are two complementary rhetorical moves of interest here. First, the appeal to clinical proof is provided, though this proof
comes in the form of a simple assertion, as if the mere utterance of the
claim were enough to substantiate it. This is hardly scientific, and what is
interesting is the irony of such an unscientific statement camouflaged as
scientific evidence. As a measure to counter this fallacy—in the event that
anyone should catch it—the statistics provide an added level of documentation, though where they come from exactly or what they are designed
to represent is not clear. Do the numbers represent actual headaches experienced by sufferers, in which case the tenth headache was unaffected
by using Bromo-Seltzer, or does it mean the tenth headache was caused
by something other than stress? How is stress being defined, anyway?
Nearly any activity that involves physical exertion (or social interaction,

or even common daily activities) can be stressful. Do these activities always cause headaches? Do headaches come in different forms? What about sinus headaches? Migraine headaches? Muscular headaches? Tension headaches? Is the literature referred to in a medical journal? Is it part of a larger study examining the occurrence of stress in general, or was it conducted by corporate scientists, predisposed to report a particular outcome? Where can the consumer get such information? Who are these headache specialists and neurologists, and what is the larger context of their work? These issues are deliberately suppressed, of course, since they would be merely adjacent to the purpose of the advertisement, they would take up too much space, and the answers might be dissuasive to the intended purpose: to get people to buy Bromo-Seltzer.

The second rhetorical move is almost an addendum, an acquiescing afterthought: "as most headache sufferers have known all along." This is certainly a stroking mechanism, but it is also an appeal to common sense, reason, and logic, and it helps include the consumer in the authenticity of the findings. Not only would any reasonable person accept these medical facts, but the advertisers have created a participatory venue in which the consumer can associate with the scientist, if only on a level of vicarious agreement. It doesn't matter that the doctors and nurses aren't real, in the sense that we have qualifying credentials with which to judge the validity of their expertise. What matters is that the readers associate the medical professionalism with their own good common sense, their faith in scientific truth and the relationship between the findings about stress headaches and the solution that is Bromo-Seltzer, a product endorsed, or so it seems, by the medical and scientific community. This is a classic example of how advertisers create a loosely woven tissue of ideological leanings, symbolic suggestion, imagery, and social ritual to erect an effective icon for the purpose of selling a product.

Again, such advertisements mirror the social expectations and orientations prevalent at the time, but they also help shape public perception of the world. This created a reciprocal relationship where the press (and other media) and the public were continually reshaping and reevaluating the events and attitudes that governed social life. Media images fed off of public expectations, but these expectations were just as often fostered by the media itself, sustained by myth and assumption, history and prospects, cultural bearing and collective association. Where magazines like *Life* were ostensibly giving the people what they wanted, they were also directing—or at least guiding—what, in fact, it was that they wanted. And what they wanted was a sense of reassurance and a pervasive feeling that the good life was not an illusion but could persevere through any crisis, even one as daunting as global annihilation. In order to provide this sense of well-being or, at least, to fend off total panic, society needed the self-assurance, authority, and expertise that science provided. It is interesting that, as in the case of the Bromo-Seltzer advertisement, the manifestation of reassurance through science was not always massive and all-encompassing; rather, it found its way into the mundane, everyday items and issues that resided in American life. The effect was totalizing, a sense that if the little things could be kept under control, the things encountered everyday in the routine pursuit of the good life—that the details were handled—then the larger issues were also well under the control of the iconographic scientific/governmental authority.

As a supplement to the reassuring role played by *Life*, and like other popular journals of the time, the magazine routinely showcased the triumphs of science's problem-solving capacities. But rather than stroke the scientific community directly, *Life* ran features that demonstrated how advancements in technology and accomplishments in engineering affected the surrounding and extended community. One example is from the December 22, 1952, issue, in which the magazine ran a story entitled "The Steel-Hungry Nation Gets a Mighty New Mill." Here, the author describes the enormity of the Pennsylvania project through pictures and text, putting the implication in perspective:

> Only 20-odd months ago the first bulldozers moved in on the truck farms along the Delaware River where the steel mill was to rise. Less than a year ago [reference to photo] the worst of the earthmoving was done and frameworks were rearing up. Today the plant is the heart of a huge industrial complex [reference to photo] with docks for ore boats, and 75 miles of railroad. Around it, to house the 6000 workers, new houses are going up. Fairless Hills, the first to get started, has already built 1100 and mass-builders Levitt & Sons have begun 16,000 more. On top of this, scores of new plants— among them a huge Chrysler tank plant—are moving into the area to supply Fairless or to buy its wares. Surveying all this at last week's opening ceremonies Board Chairman Fairless, for whom the works are named, made no

mention of the $450 million the plant had cost. Instead he called attention
to the economic system which had built it. "This plant," he said, "has been
built to serve a great national need. . . . It has been done by private industry
at private risk; and every dollar that has gone into it has been our own." (63)

Note that while there is no direct mention of the role science played
in this enterprise, the influence of science is just beneath the surface in
the technology required, in the modern engineering methods necessary,
and in the development of the surrounding community. This is one example of many to suggest the rising demand and the subsequent supply
orchestrated in the interest of providing a steel plant to meet the rising
material needs of the country. In effect, the mill touched everything
around it, from the new houses going up, to the new industry peripheral
to the plant, to the entrepreneurial spirit that drove the endeavor in the
first place. The numbers help give the reader a sense of proportion, even
by today's standards, and the suggestion is that this is a world-class
plant that will not only help bolster the economy of the immediate community, but also make the country stronger, richer, and more competitive on a global scale.

Specifically, the passage above illustrates a common trend of the postwar years: the opportunity driven by the need to demonstrate power,
wealth, and stability, an impulse generated by the anxiety people felt
about their potential powerlessness, poverty, and flux. Note the emphasis
on the number of workers employed by the plant; note the boasting of the
miles of track and road, and the number of buildings; note the celebration
of man over nature, a record in concrete and steel of human conquest over
their surroundings; note the far-reaching implications of the plant—that
it will serve the country, patriotically, economically, and materially; note
that of the scores of plants that have been erected around the steel mill,
only the Chrysler tank plant is mentioned by name, a testament to national security and military strength; note the way that communities
begin to pop up all around the plant because it is more than a steel mill—
it is a means to the American Dream; note the price tag, which draws attention to the depths of private American financial reserves; and note the
comment of Chairman Fairless, who applauds the private risk taken in
the interest of furthering nationalistic industry (a minor risk it must have
been, too, considering the demand for steel in a modern, industrialized na-

tion intent on establishing itself as the major world power of the twentieth century). If the writers of this article were to compare their style, language, and presentation to that of, say, *Pravda*, they would be amazed (and shocked, no doubt, if they could get past their own denial) at the similarities. This is a patriotic document, an exhibition of the American will to empower itself and its people. And undergirding all of this is the technology, the science, and the engineering that made it possible. A country is only as strong, the story implies, as its ability to seize upon the resources at its disposal. Only through advanced building, mining, processing, and engineering techniques can such a project succeed, and America's will is strong to succeed because if it shows any weaknesses, its enemies will exploit them without hesitation.

Other examples abound. Only a week earlier in the December 15, 1952, issue, *Life* ran a story on the "Biggest Sphere for the Atomic Sub Engine," parading the latest advancements on the military and technological front:

> Last week the Atomic Energy Commission partially lifted its veil of secrecy and allowed *Life*'s artist to make a drawing of some details of the prototype of the second U.S. atomic submarine engine and the strange house that holds it. The building, now going up near Schenectady, N.Y. will be the world's largest man-made sphere, a $2-million, 225-foot steel shell. In the sphere General Electric scientists will build the atomic engine, encase it in a section of submarine hull, submerge the hull in a water tank, and learn how to run the engine. When they eliminate all its "bugs," they will install a duplicate engine in a seagoing sister to the submarine Nautilus, which is now being built and which will be run by an atomic engine made by Westinghouse. G.E.'s engine will consist of an atomic pile to produce heat, pipes of liquefied sodium metal to carry the heat to water boilers and create steam and a steam turbine to power the submarine's screws. Before building the engine, which conceivably might leak some radioactive gases, the scientists will make sure the big ball is leakproof by going over its 1364 welded seams with an X-ray machine. (31)

It is no accident that the AEC made the decision to lift its veil of secrecy (which, in fact, was hardly more than replacing one distorting screen for another), and it is doubtful that the decision was made in the

interest of democratic awareness-raising. It is more likely that the purpose was to deliberately leak the story so that the public could see the
strides being made in the interest of national defense, so that any communist infiltrators could get a glimpse (albeit it a deliberately provocative and incomplete one) of American technological know-how, and so
that the nervously insecure public could be placated all in one efficient
public relations package. Again, science is the anchor that holds down
all of this, and it is through stories such as these that there is a cross-
section of the issues and concerns that doggedly preoccupied the postwar mind. Like the steel-mill feature, this story emphasized the price
and the dimensions of the project, but it also made the direct connection
between science and the military and how this relationship was a primary contributor to winning the Cold War.

But how does one win a war that has no front, no physical hostilities,
and no traditional geographical, tactical, or strategic objectives? This was
no small question; whereas the United States was accustomed to fighting—
and winning—wars that required the movement of men and materiel
and the seizing of specific military targets, a Cold War had no such
boundaries or goals, and a rethinking of traditional military approaches
was required. But since there was no model on which to base the current
situation, innovative tactics needed to be developed. The primary intent,
as we can see through the popular magazines studied in this book, was
to establish a unified national front, one that projected an image of solidarity and single-mindedness, of unequivocal purpose and principle, despite the fact that genuine attitudes were far less certain and exact.
Through the *Life* stories, however, there was a concerted effort to pull the
nation together, and science was as certain and far-reaching an entity as
could be hoped for.

The two companies mentioned in this story, for example, were and
have been leaders in scientific and technological industry. General Electric was a pioneer in the development of the first operational American
jet engine, and Westinghouse was the first to produce an actual jet in
1941, although its contributions were more instrumental in the areas of
radar and the manufacturing of the uranium used in the first atomic pile.
These companies, in other words, were part of the American lexicon—
household names—and this created an odd sense of familial attachment
and security in the minds of most Americans. U.S. consumers, good capi-

talists that they are, have always had a certain loyalty to their favorite name brands, but there was more to companies like GE and Westinghouse: there was a feeling that they were instrumental in making the United States the global superpower that it was. These corporations were more than mere producers of goods; they were the backbone and spinal cord of the American way of life, the very reason that America won the war and became a prosperous, influential, and dominant country. Companies like these, for all intents and purposes, were America—they took the risks, represented the entrepreneurial spirit, created the jobs, hired the scientists, developed the technology, and applied what they learned to every facet of American existence, whether they touched domestic interests or international conditions, and it is not surprising that clear and easy association would be made between these companies and science and science and the American capacity to win the Cold War.

Examples abound of *Life* stories, features, and photo-narratives that illustrate this case further (an advertisement in the March 23, 1953, issue showcases the accomplishments of GE in radar gunsights for the navy; the March 16 issue of the same year has a feature on the benefits of atomic radiation for the preservation of food; there are countless ads for new gadgets, medicines, and technological products and services; and so on *ad nauseam*). To provide a rounded-out sense of the meaning of *Life*, here is a story from the March 30, 1953, edition on the seventy-fourth birthday of Albert Einstein. There is nothing transparently noteworthy about this piece at first glance; it seems merely to be a typical, journalistic human interest story on the most famous scientist in history, the man whose name is synonymous with the word *intelligence*. Yet the effort to humanize Einstein on the part of *Life* is itself a significant issue, since it is a clear attempt by *Life* editors to bring Einstein down to earth, to make him accessible, friendly, and unintimidating. Describing Einstein's preparation for a birthday luncheon in his honor, the author writes: "For the occasion Einstein shed his characteristic baggy sweater and slacks, put on a gray suit. But he found it less easy to shed a lifetime of shyness. In an apprehensive moment before the celebration, he mused, 'Even the ambassador from Mars will be there.' The 100 distinguished guests were all from this planet, but Einstein felt light years away from relaxation anyway. Eyeing a large slab of roast beef, he exclaimed, 'This is for lions'" (72). While Einstein was always portrayed as an amiable man, his reputation for intelligence was intimidating to most people, obviously, and this was not necessarily the desired reaction if the intent was to popularize science and its practitioners for the purpose of mollifying the community. This feature, however, attempts to equalize the myth, so to speak, by portraying Einstein as a shy, introverted but very human, ordinary man. One of the pitfalls of mythologizing someone while attempting to make that person accessible is that the very characteristics that helped perpetuate the image in the first place also serve to alienate those who embraced the myth, thereby creating a distance and a failure to identify. By letting the audience in on a few personal details about Einstein, the writers were able to establish an intimacy with the man that made him even more appealing. Similar to the fatherly image of Eisenhower, Einstein was like a brilliant but eccentric uncle who visited occasionally and happily, dropping quotable phrases and tidbits of generalized wisdom. We may not understand the special theory of relativity, but we can certainly relate to the dread we feel at formal social events, and, look, the greatest scientist of all time has these feelings too!

This approach, in fact, tells us almost nothing about the man, but the rhetorical impact seems much richer than it actually is. Einstein, in many respects, already embodied the stereotypical scientist. Any quick look at science fiction movies like *The Day the Earth Stood Still* will demonstrate this clearly. In this film, an alien from another world (a planet in our own solar system, apparently, since the alien reveals that he has traveled only 250 million miles, hardly a huge distance by galactic standards) seeks the greatest mind on earth. As one might guess, the greatest mind is a scientist, and one who looks remarkably like Einstein himself, with his short stature, herring-bone ruffled tweed, and wild hair. Even his behavior is typical of the mythologized Einstein: he is difficult to get hold of, he is sequestered in his study, complete with a chalkboard full of baffling mathematical calculations (scientists apparently can only think while writing on chalkboards; pencil and paper is not an alternative), and a funny little German accent. The portrayal is no accident of casting and costuming; the scientist must be a recognizable icon the audience can believe is really the world's greatest mind. And because of this recognizability, we feel comfortable in this presumption. However, being the world's greatest mind must be an overwhelming responsibility, for both the man and the man's followers, so a little levity and humanization helps bridge the gulf between scientist and John Doe, thus making the leadership role sought for science more palatable and the public more responsive.

Life's function as a cultural document, then, served a number of important purposes. Its very mission was to capture the American consciousness in pictures and words, and it clearly reflected the public need for reassurance and security through scientific activity. In the Bromo-Seltzer advertisement, we see the culturally ingrained acceptance of science as the leading authority not only for state-of-the-art high-tech weapons and production, but also for ordinary household items. The practice of using science, or at least the icon of science, as an authority for peddling consumer goods is ubiquitous today, and we hardly notice the rhetorical import behind this approach, but during the 1950s it is especially clear that the appeal to science was the ultimate appeal to the consumer's sense of logic, common sense, and stability. If scientists endorsed a product, it was reassuring indeed to know that the best minds in the country were behind your purchase. Advertisers were well aware of the cultural barometer that indicated a support and even reverence for science, and they exploited this effectively in advertisements that appeared in major popular journals like *Life*. More than simply mirroring the social climate, however, such journals helped shape and perpetuate the public mind-set by seizing upon the cultural predisposition toward scientific activity, producing a reciprocal energy that created social acceptance while simultaneously taking advantage of it.

Another important feature of the *Life* texts was its function as a documentary highlighting America's strength and industrial capacity. As a public relations tool, this had a rhetorical effect analogous to the Soviet parades, which brandished convoys of military equipment and formations of marching soldiers in the streets of Moscow. Only this was subtler. Rather than make an ostentatious show of military might that was as much for the ego of the leaders as for the morale of the country, *Life* features served as a gentle reminder that ours was a nation that had not only the natural resources necessary for strength and stability, but also the technological know-how and engineering innovativeness to use it. And there were several happy residual effects to this line of persuasion: while it showed industrial strength, it also reinforced public hopes for a strong economy, as jobs were created and filled and peripheral communities were established, creating even more commerce and opportunity. And science maintained the status it enjoyed as the leading industrial, commercial, and technological authority since it was transparently behind the scenes developing the techniques, materials, and machinery to make all this possible.

Finally, *Life* was an effective perpetrator of the iconographic mythos and solubility ethos of the scientist through its human interest stories on eminent scientists like Albert Einstein, who was the defining personification of the scientific mind. More important, *Life* deliberately brought him within reach of the average reader by cleverly humanizing him—making him not an untouchable godlike figure whose mind was some divine mutation of human intellectual ability, but rather a familial figure whose wise guidance and eccentric ways were both charming and poignant. Einstein was transformed into a man, and this allowed the country to see him as a trustworthy and reassuring ally, someone we were glad was on our side. This image of the scientist would endure as a carefully constructed icon of the power and moral integrity of the scientist throughout the Cold War, a veneration of the scientist that has not been seriously and visibly questioned by mainstream society up to this day.

Dominant Attitudes

What was the overall attitude regarding science during the 1950s? This is difficult to answer with one sweeping statement because the very nature of the postwar climate was one of uncertainty and flux. But this condition in itself provides clues about the emotional needs the country was expressing. Because a Cold War is by its very nature a war of words, ideas, and symbols, there was a certain unease in a population that relied so heavily on the concrete, pragmatic solutions it had used so successfully in the past. The internal conflict was intellectual and emotional: anxiety and doubt were masked by competence and conviction, and the contradiction took the form of a surrealistic collage of images, myths, and ideologies. Central to all this was a necessary faith in science that had, despite its marginalized detractors, emerged unscathed from the quagmire of contradictory feeling to provide a mirage of stability, direction, and certainty.

Americans were worried about a great many things during this time, and the other sources of Cold War anxiety are frequently overlooked in favor of the defining atomic age bane of nuclear holocaust. While this is obviously a primary locus of concern, there were just as many adjacent problems that occupied the American consciousness. People did not, as is often assumed, sit about fretting about nuclear annihilation. Clearly, there was a certain subconscious impulse driving American activities, and much of it stems from anxiety about protecting the homeland from Soviet attack. But just as much motivation was derived from a need for stability and hope in the face of other, equally daunting problems, either perceived or actual. How do we feed the country? What do we do about overpopulation? How much pollution is too much? Where do we find the natural resources needed to maintain both our high standard of living and our need to stave off Russian aggressiveness? Is our technology falling behind that of our enemies? Who can answer these questions, and who can solve these problems?

Science and scientists were the obvious agents necessary for shouldering this burden, and their public portrayal took form as an amalgam of competency and innovativeness. More than that, they were the behind-the-scenes operators for nearly every new domestic development that involved industry, technology, or research—new inventions, new projects, new methods, new processes, new ideas, and new solutions to new problems fell on the doorstep of science, and it was a responsibility that was publicly portrayed as welcome and manageable. The effect was relief, however tenuous and vulnerable. The public had faith in American science, but it also harbored a typical dose of skepticism. Empirical evidence was the domain of science, and the public's need for this reflected a lesson well learned.

This irony is this: Whereas the pragmatic American community insisted on empirical documentation to satiate its anxieties, it was willing to allow science to act as a surrogate for actual evidence. In other words, while society craved practical solutions to its empirical problems, it trusted science to function as an enthymemetic substitute, a shorthand mouthpiece of solubility. Rhetorically, science had been established as the authority on everything from atomic piles to toothpaste, and it was not necessary (or particularly desirable) to practice independent critical thinking when faith in the scientific enterprise could function as its own clockwork mechanism, addressing the needs of the domestic—and even global—community. Through the serious and subtle efforts of the popular press, science's unfettered role as the public's savior marched on, more influential and dominant than ever. It should come as no surprise, given the form in which the Cold War was cast, given the historical conditions that allowed it to grow, and given the social proclivity toward practical solutions to domestic problems, that science should evolve as a cultural emblem of confidence and faith, a necessary representative of America's power and capabilities. Science was more than just a practice and methodology; it had become symbolically abstracted and packaged for public consumption, such that the images and associations attached to science became part of the cultural lexicon and cultural orientation.

Chapter 5

The Images, Metaphors, and Religious Symbolism of Science

The development of nuclear imagery can be seen in a number of prevalent archetypes, but a key source was from the mouth of governmental policy makers, their activities, and the preconceived manner in which they wanted the American public to think about and react to the international situation. The speeches of Harry Truman, Winston Churchill, and Dwight Eisenhower all cast the postwar world as one that was admittedly dubious and serious in its dimensions, but also as one that required the American public to put more and more faith in the scientific aspects of possible solutions. The history of nuclear imagery and rhetoric had always retained close ties with the science that made it possible, but the nature of the Cold War was such that continued support for science was deemed a central contingency if we were to emerge from the situation as a nation intact. The public support winnowed for science was not *wholly* informed; rather, it was based on cultivated preconceptions and misconceptions advantageous to state security and activity.

The use of relatively new media sources also made the early Cold War unique in its delivery of scientific and propagandistic rhetoric. Though public service films had been used during World War II, the key difference was that America was already involved in combat with an enemy who had physically demonstrated a desire to overtake strongholds we considered ours. While these films used rhetoric to enlist public support, the ideological game being played was far more familiar. During the

131

Cold War, the situation had changed for one drastic and monolithic reason: a new technological nightmare known as the atomic bomb. It seemed
that the only entity that could deliver us from this beast was science itself.
Though many of the images and rhetoric that surfaced in the description
of scientific enterprises were not exactly deliberate, they were the result
of familiar religious and ideological symbols that had found their way
into our culture and were therefore an efficacious means for furthering scientific activities and allegiance. The solubility ethos was never more authoritative than it was with the increasing tensions surrounding the
development of nuclear weapons, but the technical details of their production were far too advanced for the average layperson. Science and scientific advocates needed a more accessible means of conveying scientific
operations and technological solutions to the public, and the answer to
this problem was not so much educational as it was rhetorical. The tropes
we normally think of as existing only in the lofty sphere of literary analysis were adopted to fulfill this need—tropes like metaphor, metonymy, and
synecdoche. These literary/rhetorical devices function as a shorthand for
more complex ideas and designs, and it is through these tropes the public
was allowed restricted access to the inner workings of the scientific activity that informed domestic and foreign policy.

The dynamic between sign (i.e., the metaphors used to encapsulate
scientific and ideological concepts) and signifier (i.e., the conditions or
ideas the metaphors refer to) is a complicated one. The intent was to educate people on the *issues* that informed scientific activity, but in as efficient
and effectual a manner as possible. The purpose was not to disseminate
widespread, detailed knowledge of science; it was to make the public
aware of the advancements, experiments, and overall condition of the nation's scientific institution. The attempt was both to rally support for
America's scientific interests and to create a nationalistic cohesion, to
strengthen the country materially and ideologically through the lens of
science. While there were nationwide programs designed to recruit tal-
ented people into scientific fields, this was a time-consuming and limited
enterprise aimed at certain promising (and often privileged) segments of
American society. For the rest of the country, a more comprehensive and
basic campaign was mounted to bring people in alignment with the nation's scientific interests. Metaphors, in particular, proved to be a useful
means of boiling down the complexities of the international situation and

the scientific solutions being pursued to address it. The shortcut to aware-ness that metaphors provided meant more people could understand—on a limited level—the true scope of the national and scientific condition. Metaphors acted, as they do in most applications, as quick, effective points of reference that nearly everyone could understand, and this in turn made collective reasoning about the state of America's scientific and ideological home front more uniform.

Theory of Dominant Metaphors

Metaphor is perhaps the most prevalent form of conventional expression at our disposal. As a linguistic vehicle, metaphor is so versatile and so use-ful as a means of conceptualizing the unfamiliar that it is almost second nature to call upon it to make comparisons, but this is by no means the extent to which it is used. Far from being merely a manner in which to draw parallels between that which is familiar and that which is not, meta-phors are used not only to clarify but also to function as a shorthand for ideological assumptions, opinions, and orientations. Many metaphors are cliche, and many of those cliches have a nationalistic, political, or even patriotic etymology, a convenient way of capturing a sentiment or an ethic without having to rationalize an idea or directive. In this respect, metaphors are enthymemetic; that is, they operate almost syllogistically, with a number of premises preceding a logical conclusion, except that many of the most basic premises are missing because they are so well accepted that it is not necessary to reconstruct them. When a man insists that his lazy stepson needs to "put his shoulder to the wheel," for example, it is not usually necessary to explain the origin of or the meaning behind this platitude. It is clear that he means the boy needs to work harder, and one can even extrapolate from this—if one were to stop and think about it for a moment—that he embraces a standard American work ethic that frowns upon sloth and rewards hard work and commitment to a project or a job well done. A lot is going on in this simple and common cliche, yet when we dismantle some of its structure, we see that it can function as a measure of cultural ideals and social expectations. A metaphor is embed-ded in the culture when it is used frequently as a cliche, of course, but there are many more subtle, ubiquitous ways that metaphors are used.

George Lakoff and Mark Johnson, in their 1980 study *Metaphors We Live By,* have suggested that metaphors are central not only to language structure but to one's very conceptualization of the world. Contrary to traditional viewpoints that teach metaphors as rhetorical tropes or linguistic conveniences that only poets, essayists, and professional writers use, metaphors are pervasive in everyday speech and even casual conversation. School lessons that taught metaphors as linguistic baubles, as merely colorful embellishments, were as reductionistic as they were erroneous because metaphors are used every day by everybody in nearly any discursive situation, formal or informal, written or spoken. Metaphors are so pervasive, in fact, that it would be hard to communicate effectively without them; people rely on them to convey some of their most basic assumptions and ideologies.

In American society, as in all societies, there are a number of key metaphors that dominate individuals' thinking and direct their conceptual perspective. These metaphors provide a locus of cultural understanding from which many other metaphorical expressions are derived. These dominant metaphors are not just everywhere in the language we use; they help govern the way we think about and act upon certain concepts. An example used by Lakoff and Johnson is the uniquely Western notion that ARGUMENT IS WAR, a battle that is won or lost, with a victor and a vanquished. Language concerning argument bears this out. Militaristic phrases are common in everyday references to the activity and discussion of argumentation, but they are more than just handy modes of expression. It is clear that people actually *conceptualize* argument as a battle, a fight that is either won or lost—we think of it in these terms, and our language reflects our mode of thinking. Lakoff and Johnson point out that

> we don't just *talk* about argument in terms of war. We can actually win or
> lose arguments. We see the person we are arguing with as an opponent. We
> attack his position and defend our own. We gain and lose ground. We plan
> and use strategies. If we find a position indefensible, we can abandon it and
> take a new line of attack. Many of the things we *do* in arguing are partially
> structured by the concept of war. Though there is no physical battle, there
> is a verbal battle, and the structure of an argument—attack, defense,
> counterattack, etc.—reflects this. It is in this sense that the ARGUMENT IS

WAR metaphor is one that we live by in this culture; it structures the actions we perform in arguing. (4)

The idea that metaphors dominate our language and that dominant
metaphors guide our orientations about everything from current events
to our own cultural/historical heritage to our outlook on future events
means that we can examine omnipresent metaphors during any given
time frame for clues about mainstream attitudes and assumptions. This,
in turn, gives us a better picture of the American collective ideology during the Cold War, especially in the metaphors regarding not only scientific ideas but also in the scientific/technological metaphors that came
to be used as touchstones for the overall disposition toward and approach to our daunting domestic and global situation.

Key Metaphors of the Cold War

Perhaps one of the best-known metaphors dominating the early Cold War
was Winston Churchill's famous "Iron Curtain" speech. For whatever
reason, the image of an impenetrable partition seemed to capture the paranoia and outright fear held by—if not yet the American public at large—
the governmental bodies that wished to impress upon us the scope, nature,
and seriousness of the Soviet threat. It is interesting that later on, after the
metaphor had had time to become part of the cultural lexicon, a physical
Iron Curtain was erected in the form of the Berlin Wall. Whether this was
a deliberate move by the Soviet government to create fear and intimidate
the Western world through the use of an overriding metaphor, or whether
it was simply a practical method of containing its citizens is difficult to say,
but the distinction is ultimately irrelevant. The Western world, especially
the United States, viewed it as a prophecy fulfilled, and this provided adequate rationale for the national policy it would maintain.

Even though the Iron Curtain speech was delivered by an English-
man, it came to represent American feelings regarding the fear of Soviet
aggression. Churchill delivered the speech at Westminster College in Ful-
ton, Missouri, in March 1946, and the passage that gives the address its
name employed a metaphor that stuck in the minds of Americans and
gave literary substance to an otherwise purely abstract feeling of dread:

> From Stettin in the Baltic to the Trieste in the Adriatic, an iron curtain has
> descended across the Continent. Behind that line lie all the capitals of the
> ancient states of central and eastern Europe, Warsaw, Berlin, Prague, Vi-
> enna, Budapest, Belgrade, Bucharest, and Sophia, all these famous cities and
> the populations around them lie in the Soviet sphere and all are subject in
> one form or another, not only to Soviet influence but to a very high and in-
> creasing measure of control from Moscow. (qtd. in Siracusa 208)

With this single metaphor, a classic of Churchillian prose, the image of
the Iron Curtain was forever impressed on the minds of the American
public. This image is one of despotic strength, darkness, and calculated
control and containment, a carefully engineered symbol of Soviet intentions. In American propaganda films, the image was given even more
substance through animated maps that showed how this Iron Curtain
looked on a geographic plane. Heavy music and a narrator's stern warning underscored the seriousness of this global situation: if the Soviets
were allowed diplomatic conquest of these satellite states, all of Europe
could fall under Communist control. There were even films that showed
what Communism might look like in a peaceful Wisconsin town, with
the seizure of free presses, the burning of books, and the imprisonment
of clergy and city officials, thus fueling paranoia in the American public
that Communist rule was possible at home. Such films were reminiscent
of the Frank Capra "Why We Fight" series of World War II, in which
clever animation and a do-or-die tone were used to enlist public support
for military operations. The films were clearly slanted, but, more important, they reflected a quiescent public consciousness, one that required
only carefully selected visual imagery to give it shape and meaning.
      Of the speech itself, close analysis shows why the words were so ef-
fective. In the above passage, there is epic cataloging of capitals, many of
which were familiar because of the European theater of World War II.
Since the public had just sacrificed the past four years and thousands of
American lives liberating these cities from Nazi aggression, it is obvious
that Churchill wished to remind the audience that these sacrifices should
not be squandered. As with all effective rhetoric, the speaker was able to
tap into the shapeless emotional urges of the audience and give them a
specific form. The reality of the situation, in other words, was a culmination of symbolic representation provided by Churchill, a means of giving

shape through verbal expression, which is easy to mistake for the way things really are. In Burke's words, "Our presence in a room is immediate, but the room's relation to our country as a nation, and beyond that, to international relations and cosmic relations, dissolves into a web of ideas and images that reach through our senses only insofar as the symbol systems that report on them are seen or heard" (*Language as Symbolic Action* 48). It probably took little prodding on Churchill's part to convince his audience of the dangers of Soviet expansion, since many of these feelings already existed in a less refined form in the public consciousness.

Even more telling is the proposal that unfolds from Churchill's speech, for it plants the seed of military upsizing in a way that seems prudent, reasonable, and sound because it draws on the personal and collective experiences in the previous war and because it is promoted by a man most would trust as a spokesman for those experiences: "From what I have seen of our Russian friends and allies during the war, I am convinced that there is nothing they admire so much as strength, and there is nothing for which they have less respect than for military weakness. For that reason, the old doctrine of a balance of power is unsound. We cannot afford, if we can help it, to work on narrow margins, offering temptations to a trial of strength" (qtd. in Siracusa 208).

The message here is that Western allies must not merely keep pace with their Eastern counterparts in military capability but must surpass them to the degree that they would not consider testing Western military strength. This was one of the first such expressions of an attitude that would lead the United States and the Soviet Union to a frantic arms race that would last the remainder of the Cold War, and, ironically enough, it had an opposite effect from what Churchill intended. The balance of power ethic would, in fact, remain sound as U.S. and Soviet nuclear capability escalated at a rate that dictated self-preservation over a preemptive strike that would surely have destroyed both attacker and attacked. This knowledge, of course, is hindsight; at the time, Churchill felt that the Soviet code of strength (whether accurate or not) necessitated massive military upsizing, and this meant nuclear weapons and delivery platforms and people trained in the science of designing and manufacturing them.

In this respect, Churchill and Truman were in agreement. Truman, however, was equally concerned that the public be informed of the dangers that lay ahead—dangers that were in part created by his increasingly stiff opposition to the Soviets. Such opposition, as Henry Wallace would criticize to a degree that eventually led to his professional suicide, was ultimately self-defeating, for getting tough with the Russians guaranteed that they, too, would be forced to reciprocate the hard-line policy of the United States. Still, Americans were swayed by Churchill's rhetoric because he reminded them that they had a huge investment in maintaining a world safe for democracy. The Iron Curtain metaphor effectively captured the notion of global division, an ideological separation that Churchill managed to give geographic and symbolic proportions. However, what Churchill really wanted was to establish and nurture an Anglo-American alliance, and he was able to express this desire by forging an ideological ultimatum that Americans saw not only as commonsensical but also as political reality.

Lynn Boyd Hinds and Theodore Otto Windt Jr. explain how a rhetorical process can actually change political reality by distinguishing between three realms of word and reality: the natural world, the verbal realm, and the order of sociopolitical reality. The last of these spheres of language depends directly on the verbal word for meaning because it has no exact correlative in nature; that is, it is a strictly human construction that defines relationships. Therefore, political reality can only come into existence through persuasive language, for there is nothing in the natural world that would aid in giving it meaning. This creates problems in understanding, since people tend to think that the language they use describes the way things naturally are. The tendency reveals our susceptibility to linguistic illusion, yet such a theory of the rhetorical nature of sociopolitical reality goes far to explain the success of speeches such as Churchill's. The abstract nature of political language is easily confused with the way things *really are* because people mistake metaphors, analogies, figures of speech, synecdoche, symbols, images, etc. with the ideas they are meant to illuminate. A skillful rhetorician realizes this innate tendency to substitute reality with the devices used to describe it and will exploit it effectively for motivational purposes.

Churchill's speech was one of the core exercises in rhetoric that helped change the political reality of the early Cold War. Through it, he was able to define the new international situation as an impending crisis, though there was no single external event that could be seen to factually justify such a sweeping claim. At the center of Churchill's rhetorical sagacity was

> his division of the world into two competing antagonistic ideologies, symbolized by the iron curtain. The new condition required new words to describe this schism, and the metaphor succinctly provided that new view. The crisis Churchill proclaimed was potentially more a threat than the old prewar crisis because of atomic weapons. What was at stake was the survival of democracy itself. On the one side of the iron curtain was "Christian civilisation," representing all that was sacred to Americans and British, who shared traditions from the past and aspirations for the future. On the other side was the profane, a tyrannical dictatorship with sinister designs on the world and determined to spread a doctrine that negated all that Americans cherished. (Hinds and Windt 104)

Hinds and Windt further point out that the notion of the Iron Curtain tapped into an image that was fresh in the consciousness of the American public, namely, the existence of a wartime front. The visual separation of Europe into east-west segments that had ideological discrepancies was sufficiently similar to the front-line battles recently fought that it could supply the rhetorical suggestion of a hot conflict where none yet existed. By creating this division, Churchill was able to draw on the "dominant and subordinate rhetoric of World War II" (107), where the former suggested cooperation in battle to preserve democratic principles from a totalitarian enemy, and the latter gave that enemy the label of Communism, a label which had already been compared to Axis regimes in its despotic potential.

Oddly, the development of the Iron Curtain image was later considered a prophetic testament to Churchill's ability as a political commentator. This assumption, of course, suffers from a fallacy of false analogy: another image could have been planted and taken on similarly ample proportions. Churchill, however, had carefully chosen the Iron Curtain metaphor for its exclusively potent properties, and it was therefore not a predictive image, but one that had taken root and thrived in the fertile soil of the postwar American mind. Just as important, however, is that his authority as a prophet had been established through the metaphor, and in this way the religious symbolism of his message had much more impact.

He had suggested the "good over evil" theme in a way that outlined policy in no uncertain terms: the expansion of this evil empire must be thwarted for the good of Christian civilization and democracy, ideological principles that were analogous in the minds of many Americans.

The press provided additional impetus to Churchill's message by writing increasingly on Soviet affairs after Churchill's speech. Editorials became more critical of the Soviets, and journalistic convictions embracing anticommunism became more prevalent. Through the mere pronouncement of a bipolar division between Russia and the West, Churchill had engineered a suggestion that permanently colored the way journalists perceived the Cold War situation. Most typical of the rhetoric used was the dehumanization of the enemy, a necessary measure when establishing an ideological adversary with whom war is a distinct possibility. For example, in the words of popular radio commentator H. V. Kaltenborn, "the communist has no respect for truth, for pity, for human life, for individual dignity. The leaders in the Kremlin are ruthless revolutionaries whose objective is the Communist revolution" (Hinds and Windt 112). Where Churchill had gotten the ball rolling with the Iron Curtain image, journalists validated this interpretation as political fact by maintaining that only a merciless group of radicals could sustain the global deadlock that the Iron Curtain suggested.

The problem with the effectiveness of Churchill's speech can be seen in the way it divided official opinion on how to deal with the Soviet Union. Where the United States wanted to demonstrate to the world that its intentions were peaceful, the hard-line measures that were taken often contradicted this position. The most famous critic of U.S. policy at the time was Henry Wallace, who was secretary of commerce in 1946. Wallace was particularly concerned with how U.S. activities appeared to the rest of the world, and while his complaints were valid, his views were becoming exceedingly unpopular. In his letter to Truman of July 23, 1946, Wallace stated clearly his view that U.S. military upsizing would be seen by the world as diplomatic hypocrisy: "I mean by actions the concrete things like $13 billion for the War and Navy Departments, the Bikini tests of the atomic bomb and continued production of bombs, the plans to arm Latin America with our weapons, production of B-29's and the planned production of B-36's, and the efforts to secure air bases spread over half the globe from which the other half of the globe can be bombed. I cannot but feel that these actions must make it look to the rest of the world as if we were only paying lip service to peace at the conference table" (qtd. in Hinds and Windt 116). For provocative observations such as these, Henry Wallace was eventually fired. But Wallace's real undoing was brought about by a speech given in Madison Square Garden on September 12, 1946, in which he endorsed a hands-off policy regarding the political affairs of Eastern Europe and suggested the possibility of friendly relations with the Soviet Union. While many regarded this as a formal statement of Wallace's alleged pacifist tendencies, Wallace denied such a position. Moreover, the speech was a direct challenge to Churchill's Iron Curtain interpretation of Soviet activities (none of which were ever explicitly outlined in Churchill's speech at Fulton), and the day following the speech, many newspapers called for Wallace's dismissal. Political pressure mounted, and by late July, Truman asked for Wallace's letter of resignation.

The importance of this event can be seen both in Truman's claim that he doubted whether Wallace's view was intellectually sound and in a *Time* magazine cover picture of Wallace featuring the caption "America Must Choose." This same magazine contained an article that heartily attacked Wallace's ideas and his counterfeit view of reality. A *San Francisco Chronicle* article made the same charge. Clearly, a line had been drawn in the sand—on one side was the Iron Curtain mentality of the American majority and, on the other, Wallace's minority that challenged such an interpretation of political reality. According to Hinds and Windt, the chronology of events following the Iron Curtain speech suggests that the symbol spun by Churchill became a fixed part of American mentality regarding perception of the Soviet Union, its activities, and American reactions to these things. The speech marked, and in some respects ushered in, a new view of the global situation that was represented by an increasing American suspicion of communistic activities and of those who questioned U.S. foreign policy designed to counteract its sinister influence.

But what effect did this have on the image of the bomb and its significance as a driving force for combating the spread of Communism? Why would the American public, who in one respect were very anxious over the prospects of nuclear warfare, support a get tough policy against the Soviet Union that guaranteed increasing tensions and pushed the possibility of nuclear war closer to realization? Part of the answer lies with the knowledge that in 1946 the United States was still the only major world power to possess the bomb, even though U.S. policy makers were well aware of the imminent development of a Soviet nuclear weapon (though this knowledge may have been deliberately withheld from, or at least diluted for, the general public—and for obvious reasons). The media also tended to cultivate a feeling of security by suggesting that if a nuclear conflict were to arise, the United States had the know-how, head start, and technological superiority to emerge victorious. Nevertheless, during the late forties public preoccupation leaned as much toward the desire to flush out and subvert Communism as it did toward concern over an atomic war. The emphasis for the moment seemed to be more ideological than technological.

The ideological considerations were as much in the forefront in the minds of scientists as they were in the mind of the public; what did the scientific community think of U.S. policy regarding nuclear weapons and its relations with the Soviet Union? This is an important consideration since science was, of necessity, at the very heart of this debate. According to Robert Gilpin, the majority of politically active scientists (though it is unclear what proportion of the whole scientific community this comprised) felt that containment of Soviet expansion had to take precedence over international control of atomic energy. This established, in Gilpin's nomenclature, two main schools of scientists with regard to nuclear power and weaponry: the containment school and the control school (67).

The containment school consisted of scientists central to the initial development of the atomic bomb, men like Robert Oppenheimer, Edward Teller, Arthur Compton, Isador Rabi, James Conant, and Hans Bethe, who maintained that the Cold War had originated with the aggressive Soviet policies motivated by nationalism and a desire for global domination. These men believed that the issue of Europe's future had provided the incentive for the arms race and the Cold War and were strong advocates of increasing U.S. attention to the economic, military, and social reconstruction of postwar Europe. Europe's fate was, of course, at the core of U.S. foreign policy; it was the concern for Europe's future that gave Churchill his influential metaphor and, once again, reminded the American public that their sacrifices in Europe during the last war should not be in vain. Also, many of the scientists who were advocates of the containment school were European exiles themselves, so it should be of no surprise that their interest in Europe's future was a genuine, if a less than objective, one.

Moreover, the attitudes of the containment school would have a strong influence over both the leaders and the public at large regarding nuclear policy because these men had been, in some cases from the beginning, at the center of the atomic bomb issue.

It is clear that the Iron Curtain metaphor had had a significant impact on the consciousness of scientists as well. Oppenheimer, for example, went so far as to advise that the United States should abandon the United Nations negotiations on atomic energy control in favor of the containment policy because he "felt certain that, if the Iron Curtain was not lifted, any plan of international control would be exceedingly dangerous to the United States" (Gilpin 71). He feared that if the UN negotiations continued, the United States would be placed in the dangerous position of securing self-imposed sanctions against the production of nuclear weapons without ever having addressed or lifted the Iron Curtain in Europe, leaving the Soviets free to pursue their expansionistic designs with impunity.

This debate between the containment school and the control school would heat up, however, as Russian nuclear testing increased and their construction of the bomb became inevitable. Another development that complicated the issue was the imminent production of the hydrogen bomb, also known at the time as the Superbomb. This development would address not the problems of war per se, but rather the problems of deterrence, a concept that was new to the minds of strategists. In short, the decision to devise the hydrogen bomb was one that was informed by the necessity, it was held, to *avoid* a war that might occur using less powerful, more conventional atomic weapons. Thus arose the need for defense intellectuals or strategic analysts whose job, under the dictates of deterrence, was to consider ideas rather than actions (Moss 237). The objective of the defense intellectual is to ensure that in the event of nuclear war some rational control was exercised over it.

This made the H-bomb, even more so than its progenitor, a rhetorical instrument rather than a practical one. Public support for the development of the new weapon was widespread because people felt that it was the only way to stave off the Communist menace that was threatening the American way of life. It was also widely felt that if we did not complete construction on the H-bomb, the Soviets surely would, putting us at a decided disadvantage in the arms race (and, it followed, increasing the possibility of Communist aggression or even a full-scale nuclear

attack). The *idea* behind the H-bomb, its destructive force, and its implications for international relations was considerably more real in early to
mid-fifties than its physical presence. That is, while the bomb had been
part of the American arsenal since the test explosion on the island of Elugelab on November 1, 1952, the results of that test—the island of Elugelab
was gone, millions of gallons of water had instantly vaporized, and there
was a crater in the ocean floor a mile wide and two miles deep (Moss
61)—proved that this was a weapon that could not be used. Therefore, the
rhetoric that drove the campaign to create the H-bomb was one based on
the assumption that it was somehow a necessary evil that the United
States must acquire before its enemies did, but one that no one had any
realistic intention of using.

The logic driving this, at the time, seemed inescapable. President Truman, in his last state of the union message to Congress in January 1953
(just before Eisenhower took office), declared the following: "From now on,
Man moves into a new era of destructive power, capable of creating explosions of a new order of magnitude. The war of the future would be one
in which Man could extinguish millions of lives at one blow, wipe out the
cultural achievements of the past, and destroy the very structure of civilization. Such a war is not a possible policy for rational men" (qtd. in Moss
62). This message appears to contain a warning to the new administration
to be extremely careful in its policy-making decisions when dealing with
the Soviet Union. While a war using such weapons was not a possible policy, what did become the grimly ironic policy was that in order to prevent
the war, such weapons must exist. This policy came into being because
the Russians were suspected of testing their own H-bomb, a fear that
would be realized in August 1953. The international situation was desperate and people were genuinely frightened.

Interestingly, the use of the word *bomb* also started to disappear from
official descriptions of the weapon. Eisenhower, upon confirming the detonation of the bomb at Elugelab, said that the test was "the first full-scale
thermonuclear explosion in history" (62). It was clear that the image of
the bomb must be eradicated from the public mind, since it could not be
used as the weapon that the *idea* of the bomb connoted: "It was an array
of delicate equipment that may have weighed as much as sixty-five tons,
built around an atom bomb, more like a whole laboratory" (62). More
than just an image of the scientific laboratory in miniature, however, the

bomb was a ubiquitous presence, a symbolic abstraction of the actual device that represented nothing short of the apocalypse itself. The attempt to soften this impression was unsuccessful, however, since newspapers usually spoke of the American hydrogen bomb, suggesting that while officials wanted to euphemize the description of the weapon, or even redirect the way in which Americans thought about it (i.e., as a deterrent, not as a weapon of war), the image of the monolithic Bomb was forever etched in the public consciousness, and this time, with greater dread than ever.

The news of the thermonuclear fusion device and the test explosion at Elugelab had leaders rushing to intensify the same sorts of safeguards that had been discussed with the earlier atomic fission bombs. Such efforts included stepping up security, more negotiations for international controls, limits on scientific research, civil defense exercises, and the construction of fleets of warplanes (Weart 156). All of these measures can be seen as a reaction to the worst fears that nuclear imagery had provoked over the past couple of decades, creating a cumulative effect in the way that America would look at nuclear weapons and warfare in the years to come. The grave news of the Soviets' development and testing of their own thermonuclear fusion bomb solidified these fears in a way that would escalate the already serious tensions between the United States and the Soviet Union.

The array of images and associations built around the Iron Curtain metaphor were central to the public understanding of international affairs. With the single image of an impenetrable partition between East and West, Churchill had managed to construct a central symbol that defined early Cold War policy. That this image was mistaken for political reality speaks to the power of one synecdochal act of rhetoric: like the symbol of the mushroom cloud, the symbol of a wall encircling eastern Europe focused public thinking in a way that allowed for a whole series of adjacent images and connected ideas to emerge. Had another symbol been used in place of the Iron Curtain, international affairs during the early Cold War period might have been very different indeed.

So in one sense, the entire basis for public understanding regarding the gravity of Soviet/U.S. tensions was defined by a single metaphor. The metaphor would also be the basis for military intervention in Korea and, later, Vietnam, because U.S. leaders felt that the Iron Curtain was real and that it must not be allowed to envelope any more of Europe or Asia than it already had. It is a grave testimony to the power of a metaphorical image that policy should be fundamentally based on such a seemingly simple rhetorical trope. The scientific dimension of the Iron Curtain image can be seen in its influence on scientists, especially in the way it urged them to rush development of a thermonuclear weapon, but, perhaps more important, it created a general feeling of urgency, a pressing need to race the Soviets technologically, to remain one step ahead of a challenging and committed adversary. Had Churchill used a different metaphor, perhaps one that emphasized not containment but cooperation—one that had not painted the picture of an impenetrable and expanding front, an image of dialogue instead of a steely gauntlet—had he used the metaphor of an organism requiring growth and careful attention instead of a sweeping curtain requiring arrest, American policy would have adjusted to the metaphor accordingly.

It is difficult to think of an alternative metaphor for the early Cold War situation, but not because the Iron Curtain image more closely matches reality than the organism metaphor. In retrospect, the Iron Curtain seems like a brilliant representation of the Soviet/Allied dichotomy, but such a reading seems fallacious. It is instead a testament to the power of a single image, much like the domino effect that statesmen used to argue for military intervention in Asia. The power of the Iron Curtain metaphor was so pervasive that it spread to all corners of American culture, such that it was difficult to think of the situation in any other way. Even today the historical orientation includes the notion of the Iron Curtain in such a way that students of American history are implicated in the orientation. It serves as a method of understanding, but it also serves as a means of limiting one's perspective. It is a terministic screen in its purest sense: by superimposing the notion of the Iron Curtain onto our way of seeing the past, it becomes difficult to imagine another possibility. Any metaphor Churchill might have used would determine the tone of the political and technological climate; it just so happens that *this* metaphor tapped into concerns and convictions that the American people were more predisposed to accept as true. The craving for an ideological enemy was great enough that the metaphor was definitive, even if that definition helped fan the flames of discord in a dangerous and rhetorically commanding way.

The Convergent Manifestation of the Trinity Test and Hiroshima Bombings

Another interesting and largely overlooked cultural phenomenon that was gradually developing during the early atomic age was the idea that science in many ways functioned as a surrogate for religion. While institutionalized religion proper still played a role in American consciousness, in problems that had a technological basis, the popular scientist seemed a more trustworthy (and pragmatic) evangelist. The guidance that defined religion in the traditional sense was in many ways being absorbed by the popular scientist, and the language that scientists used reflected this emerging role. Perhaps the most familiar and oft-quoted verse to be associated with the beginning of the atomic age, for example, is this, reportedly uttered by Robert Oppenheimer, from the ancient and sacred Hindu *Bhagavad Gita*:

> If the radiance of a thousand suns
> were to burst into the sky,
> that would be like the
> splendor of the Mighty One . . .

Upon witnessing the mushroom cloud that followed the explosion at the Los Alamos test site on July 16, 1945, Oppenheimer recalled another line from the same text: "I am become Death, the Shatterer of Worlds." These passages were used to describe the Trinity Test, the first atomic explosion to be initiated by humans, and they fostered a new attitude for a new era. In less than a month after the success of the Trinity experiment, the unprecedented power of nuclear weaponry would be unleashed on a human target in the now-infamous explosion at Hiroshima. But Oppenheimer's alleged response to the Trinity Test explosion set another precedent: the immediate understanding that human beings had captured a godlike power that required equally godlike accountability. It is no surprise, therefore, that the Trinity Test, itself ostensibly named by Oppenheimer for the "three-personed God" of John Donne's sonnet, would foster religious imagery in an effort to describe the unimaginable ramifications of such a weapon. It is this merging of religion and science in the form of a tangible device/symbol like Trinity that is referred to here as Convergent Manifestation.

Convergent Manifestation might be viewed as analogous to Hegel's dialectic in the sense that it juxtaposes—even combines—two seemingly competitive hegemonies (in this case, religion and science) that emerge as a synthesized, symbolic ideology. The difference between the Convergent Manifestation and the Hegelian dialectic is that the ideology is driven by something purely physical (in this case, an atomic weapon) that, out of necessity for human survival, must remain purely symbolic. Nuclear devices of war are technological ends in themselves that ultimately cannot be used. As a result, they take on a linguistic property rather than an operational one, and they manifest the symbolic heritage they draw forth because of their destructive function. The language available about and attitudes toward global annihilation are traditionally religious, so that after Trinity and Hiroshima we see the technological dimension converge with the religious tradition to manifest itself in the form of a symbolic nuclear icon like the Bomb.

So when we hearken back to Oppenheimer's utterance, we could dismiss it as a dramatic soliloquy, carefully calculated to create the greatest impact for the press and for posterity. Oppenheimer was, after all, known for his rhetorical acuity and had often turned well-wrought phrases for the benefit of the media and those he simply wanted to impress. However, Oppenheimer himself has been described by other scientists like Freeman Dyson as a symbol of the atomic age—a man who could balance the curious dualism of clear, concise presentation with exceedingly complex ideas. He was a man who, as the lead physicist of the Manhattan Project, became the champion of the atomic cause—a cause, like so many wars before it, that was distorted by singing the dirge of freedom. As the first spokesman of nuclear warfare, his message seemed initially plain but ultimately inscrutable; where he could argue that the development of atomic weapons was self-evident for the preservation of national interests, his anxiety over what to do with such weapons after their purpose had evaporated in the intense heat of Hiroshima and Nagasaki was much more difficult to overcome with rational argument. In his dealings with the press and the public, as we can see by his choice of verse, he resorted to religious rhetoric; he was capable of speaking with the wit and grace of a poet and the fervor of a clergyman, and he helped forge the symbols of atomic warfare that we have come to inherit.

So it is plausible to assume that the religious symbolism given to the name Trinity Test was deliberate, for it suggested a cosmic significance that Oppenheimer realized and hoped to express in appropriately deitific terms: that nuclear energy was the controlling force that bound the whole universe, and that humanity had finally unleashed its awesome power. In the search for greater weapons, scientists had the daunting task of unraveling the atom in the same impossible way that one might unravel the Father, Son, and Holy Spirit from their infinite bond. More significantly, perhaps, is a feature that linguist Paul Chilton points to: descriptions of atomic weapons drew upon cultural traditions to give the new technology meaning and make sense of its implications.

Some—for example, the Los Alamos laboratory group leader, Joseph Hirschfelder—referred to the Trinity explosion using the oxymoron *scientific-technological miracle,* suggesting that even science was subject to the impossible, to the whims and fancy of fate and the Almighty. These reflexive religious invocations indicate that while the Trinity Test created new hopes and anxieties, it also compounded old fears in the minds of scientists and leaders alike, evoking ancient dread in the Apocalypse and the darkness it would bring upon humanity. Nothing on paper could prepare these men for the miracle they had witnessed during the test. And questions arose about the moral efficacy of actually dropping the bombs on Japanese target cities, cities that had negligible strategic value and would mean the death of tens of thousands of Japanese civilians. The atomic bomb was not designed to pinpoint specific military targets and destroy them. It was designed to send a message to the Japanese government. American policy regarding Japanese fanaticism was such that American military leaders saw a need for something truly drastic, and the decision to drop the bomb was far less a means of traditional strategy than it was a rhetorical act. As Spencer Weart puts it in his book *Nuclear Fear,* "The first atomic bombing would be an act of rhetoric, a science fiction *image* aimed less at the enemy's cities than at his mind" (97; emphasis mine). Likewise, Stephen Hilgartner, Richard C. Bell, and Rory O'Connor note in *Nukespeak* that the Trinity explosion "released the metaphors of destruction that the rest of the world would not learn about for three more weeks, with the bombing of Hiroshima" (30). The first atomic explosion used in anger, therefore, would determine the way in which the bomb would and could be used in the future; it was a symbol even as it was a physical event. Where Trinity had

proven that the medium existed through which the symbol could be expressed, Hiroshima brought on the true Convergent Manifestation of
that symbol: a way to demonstrate how catastrophic this weapon could
be, and, tellingly, that after Nagasaki it would never be physically used
on human targets again. From Hiroshima and Nagasaki forward, the
atomic bomb would act as a reminder of itself to the rest of the world
rather than as an operational weapon. The image of the billowing mushroom cloud would be the powerful symbolic remnant of the weapon's
might, an image that functioned synecdochically as a warning to any
country that dared tangle with the United States, at that time the sole
possessor of atomic weapons.

Since the weapon was so terrible, it could be utilized only as a psychological and rhetorical device, for to use it as a method of all-out physical warfare was unthinkable. Furthermore, the assumption that the
Japanese were unconvinced that the United States even possessed such a
weapon suggests that we felt compelled to prove to them that we did; the
United States wanted to underscore its military resources in no uncertain
terms, and the atomic bomb was in this respect a means of pure persuasion. It was, in short, the visible embodiment of our technological ability
to win the war. Whereas the Japanese would die to the last man, woman,
and child in a conventional campaign, the realization that the atomic
bomb existed, that the United States possessed it and was willing to use
it, eventually convinced the Japanese empire that further resistance was
an act of self-imposed genocide. Even as early as the Hiroshima bomb-
ings, the bomb was quickly becoming a symbol of military might rather
than an actual weapon, a touchstone for what one country *could* do to another if it so desired. This rhetorical dimension would help define the way
the bomb would be used and understood in the future.

It is not unusual for symbolism and rhetoric to work together in this
way. The difference in the case of the Trinity and Hiroshima explosions
is in their sheer uniqueness; when symbols are invoked for persuasive
purposes, generally they are of a familiar and recognizable origin. We
automatically make cultural associations with a cross, or a flag, or an an-
them, or even a skull and crossbones or a peace sign. The atomic explosions at Los Alamos and Hiroshima, conversely, were totally new, totally
foreign, and totally terrifying, and an equally new symbolic reaction
was fostered. This symbolic dimension found an outlet in the form of re-

ligion, and the religious/atomic Convergent Manifestation that was "The Bomb" was used on a number of levels, from constructing images of Judgment Day to envisioning a twentieth-century Manifest Destiny. In fact, many scientists and scientific writers felt that science *was* the religion of the future, and that like any religion, it needed to have missionaries to spread the word. The scientific journalist William L. Laurence, for example, felt that it was his calling to take part in science as a way to counter human weakness and mortality, and this was how he represented science in nearly everything he wrote. When he was given permission to see the Manhattan Project, he felt that he had witnessed something more profound in its implications than anyone could have imagined. The project seemed to him to be "a sort of Second Coming of Christ" (Weart 100). After the Trinity Test, when scientists and leaders involved in the Manhattan Project assembled to witness the testing of their first military nuclear device, Gen. Thomas Farrell, an army engineer involved in the project, wrote that the sound of the bomb "warned of doomsday and made us feel that we puny things were blasphemous to dare tamper with the forces heretofore reserved to the Almighty" (101).

From a rhetorical standpoint, the Convergent Manifestation of religion and science did several things: religious imagery gave scientists a rationalizing mechanism with which to explain their own activities and to construct symbolic significance of their outcomes. It would also give the public a familiar orientation with which to understand the implications of this new technology—a metaphorical means of grasping what had actually occurred. While many of the images were indeed apocalyptic in scope, the piety exhibited toward science increased as a result of harnessing the new technology. Here was the beginning of a new form of a well-established attitude that embraced the notion that Americans were the chosen people, that God had selected us over our enemies for the discovery and utilization of this terrible weapon. It was a powerful rhetorical motivator, for it indicated that in our hands alone God had placed the burden of righteousness, an idea quite familiar to Americans ever since the founding of this country. Just as we had piously stabbed westward during the eighteenth and nineteenth centuries to do God's will, so were we saddled with another massive responsibility to use this weapon under the god-terms of freedom and liberty.

The Trinity Test was perhaps the first physical manifestation of what would become a common rhetorical theme after the test had taken place: science had unlocked a secret of biblical proportions, and the religious/scientific symbolism that surrounded the visual image of the first mushroom cloud would haunt both scientists and the laity. The cultural realignment that took place following the Trinity Test bears examination, since it is the precedent for the way Americans viewed atomic weaponry and our troublesome responsibility regarding it, and the rhetorical mode of expression that we relied on turned out to be both disturbing and liberating. Where many, like Laurence and Oppenheimer, felt that they had shared some profound religious experience in the Trinity Test, others felt that they had encroached upon a Promethean secret that would spell doom for humankind. These feelings were captured quite often in religious metaphor; for those who felt that nuclear power was a sign of the Apocalypse, there was still a sense that such a death could retain with it the hope of rebirth. Laurence described the Trinity Test explosion "like being present at the moment of creation when God said 'Let there be light'" (Weart 101). Where General Farrell saw the explosion as a precursor to Doomsday, Laurence thought of it as "the first cry of a newborn world" (101). In these two men, at least, we see an encapsulation of the attitudes regarding nuclear weaponry—people in general were both frightened and awed by its might and promise. Spencer Weart puts it well when he says that the Trinity Test "explosion served less to introduce new ideas than to bring ancient thoughts of apocalypse to new and vivid life" (102). The development of this rhetorical expression of science, religion, and history directed our understanding of the nuclear Pandora's box that was the Trinity Test, and constructed a cultural vocabulary that we have carried with us ever since.

The human need to invoke God when confronted with something unfathomable seems universal. One of the more capricious examples of this is the story of the GI who, having suffered a drinking binge the night before the Trinity explosion, was awakened from his alcoholic slumber at Los Alamos by the explosion and was temporarily blinded by the flash from the blast. As Lansing Lamont describes the GI's reaction in *Day of Trinity,* "the shock of the blast had induced hysteria, and the fellow lay on his cot raving about the 'wrath of the Lord' and the sin of getting drunk on the Sabbath" (243). This soldier's immediate impulse was to assign religious explanations to a man-made phenomenon. His rather

Puritan response, that the blast was a personal message from God meant to punish him for his sin, may have been erroneous and irrational, but it underscores the deeply rooted human inclination to summon religion when confronted with demonstrations of great power. It is not surprising, therefore, that religious images were easily forged out of the molten debris left by Trinity on July 16.

Consider the design of the Trinity device itself, euphemistically codenamed "The Gadget." There is a mysterious property to the schematic for this device, as if we are looking at an archaic, divine model of the universe. To the layperson, the design seems deceptively simple—everything concentric, symmetrical, and perfectly spaced. It looks like a forgotten talisman, or a pendant used in some pagan religion to ward off evil by protecting its wearer, something that Indiana Jones might be searching for. Only the words labeling the separate parts reveal its true purpose. When I first saw the drawing for "The Gadget," I was reminded of the geocentric conceptions of the cosmos like those described by Aristotle, Ptolemy, and Dante. The design is reminiscent of a time when humanity considered itself the center of the universe, the place around which all else revolved—a time before the disturbing revelations in science had tainted our rational faith and humbled us into realizing we were only one minute particle in a much, much larger scheme. It harks back to a more ancient but nonetheless logically consistent science, one that consisted of the simpler, purer notion that four elements comprised everything, and that these elements were drawn to their natural place. It is also reminiscent of a view of the universe that was spiritually consistent with Catholic doctrine, a universe where bulk material congregated at the center because it was corrupt and contaminated, furthest from the perfect, ethereal realm of God's residence in the outermost sphere. From a symbolic perspective, the exploding of this "universe" might indicate human severance from God, an attitude that suggests *we* have knowledge and power now; *we* are the creators and shatterers of worlds. Where God once held the prominent position of authority, power, and reverence, we now had stolen His design and, by extension, His command of the universe.

The Nietzschean prospect that God is dead was never more fully symbolized than in the explosion at Trinity; we had the knowledge, the power, and the will to destroy whole worlds now, a prerogative of God in the past handed over to us. From a spiritual perspective, this was very confusing and paradoxical. Was this knowledge, in fact, handed over to us by God? Did that suggest our moral righteousness in this global struggle? Harry Truman's speech following the bombing of Hiroshima and Nagasaki certainly indicates this, since God had placed the knowledge of the atom "in our hands alone," but was this merely an artful way of expressing our incorruptible intentions, or was this truly held as a sign from the Lord? It is difficult to answer this question with any real certainty, but it is clear that religious rationalization was an automatic response in dealing with this new, morally inscrutable technology. We are socially conditioned to invoke the purpose of God when we are uncertain about our actions, especially when they are so visible and so significant, and while the Truman administration did not take the decision to drop the bomb on Japanese cities lightly, having God on one's side was not a bad way to lighten the weight of moral responsibility for such a daunting and fateful decision. While this conflict and this weapon may have been human embodiments based on human squabbling and human weakness, it was God's decision to choose a victor and to provide the means for victory. Had the Japanese or the Germans developed the atomic bomb first, I have no doubt that equally religious significance would have been assigned to their cause and for many of the same reasons.

On the other hand, and herein lies the paradox, it was difficult both to justify God's advocacy *and* to suggest that we were ourselves godlike for having harnessed the atom. Had He "leveled the playing field," putting humanity on par with Himself, or were we the masters of our own destiny all along—perhaps seizing the secrets of the universe from some mysterious locale that had nothing to do with God whatsoever? It is impossible to make a generalized claim about this; however, certainly both the notion of God's endorsement and the notion of our own godlike stature were held by scientists, statesmen, soldiers, and the public. The coexistence of these conflicting outlooks is representative of the Cold War angst, since it reflected the profound confusion generated by the sheer newness of the atomic age and all its contradictions. Even scientists felt compelled to describe the issue in religious terms, despite intimate knowledge of the technical details, because the ethical/philosophical implications were well beyond the practical application of the knowledge they possessed.

Was Trinity's design the result of a residual culture subconsciously manifested by the Manhattan Project scientists in the form of a bomb to

permanently displace us from our infantile reliance on God? No. However, atomic energy was often extolled as a final technological frontier,
the last truly unknown realm of physical knowledge. Because the atom
was believed to be the most finite of particles, and nuclear energy was
the very power driving the stars, understanding it was commensurate
with understanding the building blocks of the universe. The ability to
fracture, and eventually fuse, atoms, and thus release the enormous energy within them, was considered by many scientists and military
minds to be the last great scientific discovery. With this knowledge, they
felt they had mastered the cosmos. No longer did humans need to rely on
God to make sense of the world for us; our destiny was to read the mind
of God in atomistic form and use this knowledge for our own purposes.

After the explosion, however, there seemed to be questions about
how much of this daunting responsibility we really wanted. The unprecedented devastation caused by atomic weaponry seemed to necessi-
tate religious associations once again, as though the mere existence of
such energy automatically removed man as an agent for its use. For example, Churchill made the claim reprinted in the August 7, 1945, *Times*
article that "by God's mercy British and American science outpaced all
German efforts" (Chilton 132), an early example of the suggestion that
God's will had allowed chosen nations the privilege of uncovering the
secrets of nature for their own just causes. Harry Truman made similar
statements, thanking God that He had seen fit to deliver the weapon into
our hands, and that we now had the moral obligation to do His will with
it. Human intervention and research had little bearing, it was often implied, on the development of nuclear bombs; God had *given* us the
knowledge. This is, of course, an effective way of not only removing the
human agent as the source of knowledge, but also shifting responsibility
onto an entity that is sacred, above the simplistic morals that humans
use to govern themselves. Chilton notes that the passive voice also suc-
ceeds in removing the human agent from responsibility, as in this passage from Churchill's later Iron Curtain speech: "We must indeed pray
that these awful agencies will indeed be made to conduce to peace
among the nations, and instead of wreaking measureless havoc upon the
entire globe they may become a permanent fountain of world prosperity" (132). All humanity can do, Churchill implied, is pray—the rest is up
to God "to conduce to peace," and to do this, we must rely on the Bomb's

symbolic power as a Convergent Manifestation of God's will and Western know-how and hope that the rest of the world can become as pious
as we are by recognizing the symbol for what it is. The question of moral
accountability, then, is relegated to a higher being, suggesting that our
actions with this new weapon are sanctioned by a being whose moral
compass is far more accurate than our own. Churchill more explicitly
says in an August 13, 1945, letter that "it might be claimed that through
divine grace English-speaking scientists were able to make their original
discoveries of the vast source of energy" (132). The religious association,
therefore, is taken one step further to be bound tightly to the knowledge
exposed by science. Science, though in this context not *a* religion, is
given equal status and divine endorsement—that is, the efforts of science
(and, tellingly, science practiced only by English-speaking peoples) are
approved of by the highest authority of all.

Despite our hubris and our self-congratulation, we were frightened
of the reality our new power necessitated; just as our tribal ancestors
must have genuflected to their spiritual overseers when they discovered
fire, so did we, like children who have trapped a dangerous animal and
are uncertain what to do next, defer to our creator the very power we
sought. The result was a bizarre conglomeration of science and religion,
rhetoric and symbolism, hope and fear: a Convergent Manifestation of
the spiritual and the technological that helped clarify and mystify the
significance behind this new spire that was the Bomb. It is difficult for us,
who have lived with the bomb now for over fifty-five years, to understand the strange combination of panic and rejuvenation that the atomic
bomb generated. It is only by probing the symbols seized upon and the
way they were used rhetorically that we can really understand the extreme ambivalence such a discovery created.

Machines and Man

Another contributing factor to America's scientific orientation was simple
mechanization. While machines had played a large role in American
social structure since the Industrial Revolution, the postwar years saw
industrial mechanization grow to a level of sophistication and ubiquity
that created a social landscape where skilled and semiskilled technologi-

cal workers were not only in greater demand but also in possession of technological skills that were imperative to meaningful employment. When considering the technological skills needed to operate, maintain, and repair the machines of industry, it is easy to see how the scientific paradigm was not only accepted, but deemed the only obvious orientation to adopt.

According to the September 1952 issue of *Scientific American,* some 64 percent of the entire workforce was classified as skilled and semiskilled. While these terms are not explicitly defined, *Scientific American* does provide this perspective on the matter: "In transportation and agriculture, machines by now have practically eliminated the need for human muscle power. Man has all but ceased to be a lifter and mover and become primarily a starter and stopper, a setter and assembler and repairer" (154). This is the definition of modern humanity. People no longer toil with their own backs and arms and legs to subsist; they invent, develop, and maintain machinery to do the work of many and to produce on a level that not only creates greater material output but also allows for shorter work weeks and therefore longer leisure hours—all of this thanks to advancements in industrialized machinery that can relieve humans of the drudgery, monotony, toil, and danger of working in the past. And, as mentioned in the introduction, because we enjoyed the benefits to industry of two world wars in the twentieth century, we pushed along industrialization a hundredfold from where we might have been without these conflicts. Human beings are the puppeteers in this new age of science. Where once we had bent helplessly to the whim of nature and its indifferent influence on our lives, now we had new and more powerful ways of imposing our will on it. As the great shapeshifters, we were able to extract, process, and mold raw materials into nearly anything we wished, and with the aid of our great machines, we could do so with efficiency, dispatch, and abundance.

And the labor force would not be unaffected. While unskilled labor would still be available and even needed, it would be so at an increasingly diminished capacity and it would not be satisfactorily compensated. The combination of low pay, decreased need, and diminished value placed on unskilled labor, coupled with the increasing need for workers who possessed technological skills, guaranteed the prevalence of the mechanistic model, not only in industry but also in the descriptions of social structure, medicine, and bureaucratic networks (education, for example). The dominance of the machine metaphor in so many areas of American culture dictated an undeniable point of reference for Americans. It was, simply put, a convenient and familiar way to think about the world, its promise, its problems, and its solutions. Machines had done a great deal for us. In both world wars, but especially in the second (often referred to as the first totally mechanized war), aircraft, tanks, ships, submarines, jeeps, trucks, aircraft carriers, amphibious landing craft, and a dozen other military items were central to the success of modern military strategy. Hitler's infamous blitzkrieg could not have succeeded without the trucks, dive-bombers, ground support aircraft, tanks, and transports it used. In some regards, Hitler managed to usher in mechanization on a massive scale on all fronts, civilian and military, through the use of such tactics. Mechanistic industrialization was absolutely necessary to producing the machines of war, so much so that strategic targets were almost always of an industrial nature: the oil refineries, aluminum factories, and ball bearing plants that made the war machine function. It should come as no surprise, therefore, that the benefits engendered during the war should carry over into the private sector following the war. In fact, they had never been absent from private industry. Mechanized warfare simply gave us new and more diabolical applications, but this also translated into jobs, economic prosperity, and technological advancement that could be used in times of peace, or in the case of the time frame in question, in times of cold war.

The changing face of the labor landscape had social effects as well. It is easy to see how skilled and semiskilled scientific labor would be sensitive to science and technology as a social and cultural paradigm for nearly everything that touched American lives. One reason for this, according to the September 1952 *Scientific American* article mentioned, is the alleged "uniformity of American living" and how this made the transition from unskilled to skilled labor much less visible than it might be in a more ostensibly stratified society: "In a country with a less fluid and more differentiated social structure than ours, these rapid changes in the occupational composition of the population might have brought about considerable strain. But the celebrated, and often criticized, uniformity of American society renders the effects of such transition almost imperceptible" (154). I suspect that the categorical claim that America is as uniform in 1952 as the author suggests is overstated. It is perhaps the case, based on the attitudes and assumptions regarding diversity in the early fifties, that this claim has a certain specious truth about it; nevertheless, it suggests, at least, a homogeneity of thought amongst those who dominated the press—an assumption that people were accepting of the mechanistic model of industry and social structure and that therefore such a labor force shift could take place seamlessly.

The idea that mechanization—in the workforce, in industry, in everyday social encounters—determines thought processes and susceptibility to certain orientations has a basic literary grounding. The familiarity with and exposure to the devices we daily take for granted often manifests itself in the metaphors we use. The metaphors may be used to describe the unfamiliar, to make colloquial explanations of technical or sophisticated phenomena, or simply to reorient ourselves in the event of disorientation. According to Lakoff and Johnson, metaphors contain a systematicity that aids in conceptualization, producing a pattern of thought that guides our perspective and often limits it. When we consider this alongside the idea that the labor force in American society was becoming increasingly reliant on machines and technology, that the very definition of what it meant to be a worker carried with it implications of technical skill and the use of machines, it is clear how science and technology can dominate our thinking and our approach to the world.

Variations on the machine metaphor found their way—either directly or indirectly—into a variety of social contexts during the 1950s. Even the most rustic settings in agriculture and animal husbandry had become mechanized, systematized, and industrialized, leaving a wake of mechanistic description that would become common parlance for the American farmer. In the December 15, 1952, issue of *Life,* an article entitled "Wheels Are Its Basic Tools" gives the reader a sense of the changing face of American farming. Airplanes, combines, huge tractors, and other mechanical planting, fertilizing, and harvesting equipment became part of the American farming arsenal. Considering that even as late as the 1930s agriculture consisted mainly of relatively small farms run by individual families in most of the United States, the shift to large-scale, highly technical farming is a remarkable phenomenon in such a short span of time. Farms of 1,000–5,000 acres were not uncommon in 1952 (especially in the sparsely populated Western states), and such huge plots of land required specialized equipment. Between 1920 and 1960, the average farm size more than doubled, from 148 acres to 303 acres, due largely to advances in agricultural technology that allowed single farms to work more land in a cost-efficient and manageable way. Even more interesting is that the percentage of farmers in the population's labor force decreased during the same time frame, from 27 percent in 1920 to 8.3 percent in 1960 (USDA, "Farmers and Land"). These statistics, though rough and indicative only, suggest that while the population had increased, making greater demands on the agricultural superstructure, the actual percentage of people farming had diminished. While people who once might have become farmers were going into other professional areas (many of which were technically related), fewer people were required to farm in order to meet the nation's domestic and export needs, and this could occur because machinery and advances in agricultural science had made the small farm obsolete.

Far from the sort of collectivization of farms that had taken place in the Soviet Union, the United States was developing its own brand of socialized agriculture through the use of the co-op, which allowed farmers to lease expensive machinery and share in the profits from multiple farms in a way that was nonexistent just thirty years earlier. In 1920, if a farmer was lucky and relatively successful, he might own one of the few reliable light tractors available to increase crop production; but by 1960, tractors, self-propelled combines, and trucks were not only available but necessary to the successful farm. Another telling statistic is the number of man-hours needed to produce 100 bushels (approximately four acres) of wheat: in 1920, it took nearly twenty man-hours, but by 1960 it took only five. And advancements in machinery were not the only benefits being reaped by new technology. Fertilizers, herbicides, and insecticides had all been developed and used widely by 1960. Fertilizer use alone was five times as common: in 1920, 6,845,800 tons of fertilizer was used by U.S. farmers; in 1960, 32,373,713 had been used (USDA, "Farm Machinery and Technology"). Farmers were producing far more using less land and fewer people in the space of a mere forty years. Science and technology had touched even the most outlying regions of American society, and one result was the cultural saturation of mechanization that would help define the age.

"Control and Smooth Performance"

The phenomenon of "control and smooth performance" is seen most clearly in the industrial sector. With the rise of business opportunities, information resources, and their relationship to industrial production, references to efficiency, teamwork, and capacity became the dominant phrases used to describe positive production values. Such phrases operate under the mechanistic metaphor because they suggest a functionality that is accomplished by individual parts working together for a common purpose, the reason for the machine's existence. Just as machines prevailed in the American landscape—rural and urban alike—enveloping the bodies and minds of society, creating an atmosphere of man/machine interrelationships, so did machines and their purpose become a part of our means of expression. This is a strange intersection of the human and the human-made; machines were an everyday aspect of existence that functioned as both tools and as entities in need of maintenance. This was also the time when science fiction writers and reactionaries regularly warned of the day machines would overtake us, when they became self-reliant and therefore in control. On a more symbolic level, machines were already in control: people were lost without their cars, washing machines, ovens, televisions, and radios. Just as today, the man-machine relationship was an ambiguous one; humans are in many ways at the whim of the devices that serve them. The recent Y2K scare is a contemporary example of humans' tenuous coexistence with machines, prompting survivalists to concoct conspiracy theories and amass weapons from their compounds in the woods.

This faith has its roots to a great degree in the increased mechanization of the 1950s and its consequently rising mechanistic linguistic orientation. In the January 7, 1952, issue of *Time,* for example, an advertisement for Moore Business Forms represents a fascinating causal link between the purpose of a product and the production concerns of society. The heading reads "Teamwork on wings . . . via the dotted line." The teamwork metaphor is mechanistic at its base, suggesting the coglike cooperation of various and purposeful parts, but the "on wings" reference is even more interesting. The accompanying illustration shows a Rosie the Riveter icon working diligently on the wing of an airplane, while in the background a man writes dutifully in a ledger of some sort, presumably reporting as a quality control agent or safety inspector. Above this we see three representative Moore Business Forms, one for checking, one for reporting, and one for purchasing, each of which displays a superimposed image of a flight attendant, pilot, and secretary, respectively. Dotted lines link these

forms to the activities of the first scenario described, then to a small picture catalog of the company's products: sales books, typewriter forms, carbon documents, register forms, etc.

Even to the casual reader, this requires some interpretive skill. The
message, presumably, is that the bureaucratic aspect of the airline industry requires the use of proper documentation in order for air travel to be
safe. On another level, the suggestion is even more symbolic. "Teamwork
on wings" is a provocative idea, especially when one considers the dubious notion that paperwork is an efficient means of holding together an in-
tricate unit. Anyone would concede the necessity of record-keeping; it is
a drudgery that complex institutional systems must accept in order to
function properly. When documentation is clear and concise, the mechanism runs much more smoothly. In this respect, the symbolic suggestion
is that business forms are a built-in monitoring system that assure the
smooth operation of each individual component. In a mechanistically
modeled network such as the airline industry, one can make the reasonable claim that knowing and recording the operation of each system component does in fact help ensure safety in practice. The humans who
monitor the machines that provide the service—in this case, aircraft—
must operate in an efficient and reliable manner, just like the aircraft they
service. The mechanistic mirroring here is fascinating. Each part of the
technocratic industry is an extension of the machines it uses, functioning
as an extraneous mechanism every bit as important as the internal mechanism of the machines themselves. The ad suggests the following:

> Strength in the air makes us secure on the ground. Keeping aviation fit for
> this takes teamwork, the kind of control and smooth performance turned in
> on the DOTTED LINE—the perforation on business forms made by Moore.
> Moore forms promote teamwork everywhere—in manufacture, mainte-
> nance, transportation, freight. The forms on this page make efficiency a part
> of *Checking* (3-part Flight Coupon); *Reporting* (3-part Flight Record); *Purchas-
> ing* (8-part Purchase Order). In a hundred other operations, the DOTTED
> LINE makes efficiency pay because it saves writing, simplifies handling,
> speeds work flow. *It is a FORCE that runs through an organization, keeping per-
> formance up and costs down.* On most styles of forms Moore is producing to
> capacity—its new, greater capacity—supplying industry with DOTTED
> LINE efficiency to help make America strong. (*Time,* January 7, 1952, 27)

High-tech aircraft—and automobiles, boats, trains, motorcycles—require control and smooth performance. The humans who are the starters and stoppers, the fixers and assemblers, the administrators and supervisors require an equal degree of control and performance if the machines they rely on are to work efficiently and, most important, profitably. When a society relies heavily on machines to hold together an economic superstructure, the people of that society become encompassed by the machine—in some cases, literally—and must perform on the same mechanistic level and in the same mechanical manner, a macrocosm of the technological operations that drive the social apparatus. The Moore advertisement emphasizes performance, but it also stresses control, a subtle reminder that the human element is still a necessary component of the industrial machine. Speed is also highly valued. The faster the mechanism runs, the more efficient it is, and efficiency translates into profits. Humans, unlike machines, are unpredictable and vulnerable to activities that are not in line with their specified function, however; therefore, careful documentation is needed to keep them from their erratic tendencies. It is interesting that the link between human and machine here is symbolic. Language (in the restricted, business form sense) is the cohesive element, the bureaucratic documentation that holds the synthetic and the organic together as one smooth working unit in a hundred operations. Efficiency, simplification, capacity, and speed are the key descriptors behind the metaphor.

This is only one example among many, and advertisements offer a unique perspective into dominant social attitudes because they are deliberately designed to tap into the god-terms and orientations people are most likely to respond strongly to. Advertisements function in this way as both a mirror image of the ideology and a means of guiding it in predetermined directions. Take, for example, another metaphor from a General Cable Corporation advertisement in the same issue of *Time* magazine, "A Strong American Is the Bulwark of Freedom." The god-term *freedom* is here compared to a rampart, a defensive structure associated with military operations. Such a comparison is not unusual, especially considering the time frame, but what is interesting is that the association is removed one step further by suggesting that cable—in this case used as an electrical conduit—is a symbol of freedom itself. In enthymemetic fashion, the reader is asked to deduce that the company is a patriotic force that exists solely to strengthen the American industrial front by supplying the necessary power to the manufacturing machine. Besides the company's name in the corner of the page, no other explicit endorsement of the product is provided; instead, this jingoistic paragraph appears beneath a picture of trucks, bombers, battleships, tanks, and factories:

> Our country derives its strength from its unconquerable force of fighting men on the battlefront and on the home front in government, in education, in religion, in science, in industry, in agriculture, and in business—men of integrity and energy, men with the courage to face issues, free men, Americans of all races and religions who, fighting and working together, have brought our country to its present position of world leadership. To maintain that position, to fulfill the need of free peoples throughout the world for guidance and for inspiration, to gain a lasting peace and to insure an enduring freedom, we dare not compromise the ideals of our democracy or, through complacency, surrender to the forces which oppose them! (52)

While this sounds more like a state of the union address than an advertisement for twisted strands of steel wire, it is perhaps that very intersection between the commercial product and the political ideal that is so startling. The unabashed, tireless use of the word *freedom* alone is enough for today's more jaded reader to put up his or her *own* bulwarks, but be that as it may, it is clear that the association between business and national strength is "freely" offered as a truism that would be accepted by enough people to be effective. The metaphor here is at once complicated and easily processed, and the irony of this is in many ways its strength as a rhetorical device. By creating an illusory sense of patriotism through a dubious syllogistic process, the creators of this advertisement have forged an allegiance with the American public without directly fostering their consent; they assume that the public will see the logic behind the tapestry they have woven, regardless of whether there is any intrinsic connection between steel cable and democracy, big business and national security. What is truly odd is that there does not seem to be any attempt to sell the product, at least not in any candid way. It seems more important to the company that their existence is regarded as a foregone conclusion, that it is through the ubiquitous and indefatigable presence of companies like General Cable that the country maintains its superpower status.

This industrial strength and economic security is the metaphorical bulwark mentioned in the advertisement's introduction. The dichotomy of freedom and tyranny had gone far to win the war in Europe and the Pacific, so it should not come as a surprise that the reflexive impulse to tap into the liberty ideology was a tried and proven method of garnering public support. What is unusual is the associationism generated in the interest of national coherence. In a society reliant on the production, development, and maintenance of technology, reliable means of transferring power become supremely important, so much so that the relationship between power delivery and freedom was easily forged in the American mind. The pictorial arrangement in this ad suggests that the connection between civilian and military interests is a deeply reciprocal one, where a strong industrial and economic foundation equals a strong military presence, and the conduit between these two areas of social formation, both literally and figuratively, is technology. The advertisement features the aforementioned instruments of war on the right side of the page and a city skyline with suburban houses and power lines in the forefront on the left, indicating a direct line of influence between the two entities. The metaphor of the bulwark seizes on one overarching feature: that technology is the *control of energy* and the *transformation of resources* into means of defense and military solidity.

One happy by-product of this system is that it creates a solidarity and a stability that Americans craved. Whereas the deliberately forged relationship between technology and national strength helped reassure the public that we were healthy enough to maintain a revered and carefully guarded way of life, scientists—especially doctors—were reminding us that our *individual* strength was important too. In the January 7, 1952, issue of *Time*, we read what appears to be an article on stress management: "How to Cut Your Worries." The author, Dr. Paul Popenoe, director of the American Institute of Family Relations, tells us the following:

> Many people who call on me for guidance are so confused by worry that they cannot think clearly. Intensive worry usually prevents people from being able to see the solution to a problem—even when that solution is obvious.
>
> The first step is to stop worrying aimlessly and analyze your problem. You will find that some problems are the kind you can't do anything about.

Learn to put worries of this kind firmly out of your mind. If you don't, they will make it more difficult for you to concentrate on the ones you *can* solve. (43)

This appears to be one of the self-help treatises à la Norman Vincent Peale that was gaining considerable popularity at the time and is still popular today. But it is actually an advertisement for life insurance. Next to the text is a picture of Dr. Popenoe himself, wearing a suit with lapels wide enough to be air foils, a carnation in one of them, seated in front of what appear to be book stacks in a library. Popenoe's endorsement lends credibility and authority to the service that insurance provides, of course, as does his accessible but authoritative countenance before the reader. He is the iconographic picture of the gentle, caring physician, wearing a pleasant smile and offering, in general, a reassuring presence. The setting is one of knowledge and wisdom, the place that most people would associate with education. Whenever an authority is being interviewed, either by the news media or an educational television program, he or she is placed in a setting conducive to the kind of knowledge the authority is purported to possess. In the case of a lawyer, he is interviewed in a law library; in the case of a biologist, in a lab; in the case of a professor, in an office strewn with the objects of her specialty; in the case of a geologist, amongst large rock formations. It is as if the viewer/reader is expected to believe that Dr. Popenoe has read and committed to memory every book that sits reverently behind him. He is the personification of knowledge, and we trust him for the knowledge/power we assume he possesses. He is the scientist for the average person; he has studied the data and the research and he knows what he's talking about, and he is willing to impart this knowledge to us. Dr. Popenoe is not a young man, either. He is in that happy intersection between experienced practitioner and doddering old man, an age bracket carefully chosen to elicit feelings of trust over suspicions about his competence.

Beyond the rhetorical imagery of Dr. Popenoe himself, the real point of interest here is the metaphor, "cutting your worries." Anxiety was a key motivator driving many of the attitudes and much of the policy that defined the Cold War, and the idea that one can cut out worry as one might cut the fat off a pork chop suggests that anxiety was not only a prevalent cause for concern but also an irrational, emotional state easily remedied. With the symbolic, physical, and mental health of the nation at stake, it was important that people realize that they really should not trouble themselves with issues beyond their control. Hearkening back to the serenity prayer, "God grant me the serenity to accept the things I cannot change, the courage to change the things I can, and the wisdom to know the difference," we see the validation of this conviction in the endorsement of Dr. Popenoe. In more specific terms, Dr. Popenoe suggests that we "take our worries up one at a time," another quotidian nugget of advice suggesting a quick resolution to our chronic syndrome of worry. Anxiety is not a complex network of pressures exerted on the mind by the outside forces of everyday life in the atomic age, Popenoe implies; it is the result of simple irrationality and the inability of the average mind to cope with the overwhelming complications that modern living has foisted upon us. The solution is simple: cut the fat. Take things one at a time. Analyze your problem, don't react to it. Concentrate your efforts on problems you can solve. Above all, don't expend energy thinking too much about that which is beyond your own limited scope of influence or mental capacity. Dr. Popenoe has conveniently supplied us with a metaphorical sedative by explaining a complex psychological experience in metaphorical terms that we can understand, and he has given us a method for coping with the stresses of life: "Take [your problems] up separately. Go over them calmly and carefully—and decide what you can start doing about them *immediately*. Once you actually start working to solve a problem, you'll find that you tend to stop worrying about it" (43).

We cannot leave this example without examining the true purpose behind this advice: to sell Massachusetts Mutual Life Insurance. While Popenoe's advice seems sound on its face, it is, of course, reductionistic. The legitimacy of his expertise is further compromised—at least to the discerning reader—because it is advice not used for the ultimate well-being of a patient, but for the solicitation of a service. But the metaphor of cutting out worries is a tempting one, if only because it is so simple. This metaphor is supplemented by Massachusetts Mutual itself in its own self-interested way: "Life insurance is especially helpful in overcoming worry. A well-planned program of life insurance can help to increase your *peace of mind* by making sure you will have funds available to meet specific problems when they arise" (43; emphasis mine). "Peace of mind," of course, is a mainstay metaphor of the insurance industry, and here it overlaps with Dr. Popenoe's advice in a transparent way: One method for solving specific problems is to have enough life insurance, complementing Popenoe's assertion that the best way to rationally approach an overwhelming series of stressful complications is to take them one at a time. It takes little imagination on the part of the reader to extrapolate the problem in question: What happens to my family when I die? In the 1950s nuclear family, the man of the household was typically the primary provider, at least insofar as the media tended to represent the model family unit. His death meant the absence of income, hence the absence of security for the family. Life insurance meets a specific need, providing, as an added benefit, "peace of mind." A mind at peace is a wonderful, if elusive, thing, and a "well-planned program of life insurance" adds to, if not wholly provides, this peace. Interestingly, peace is a recurring concept in Cold War mythology, as in the B-36 strategic bomber, The Peacemaker, a notion Stanley Kubrick later parodied in *Dr. Strangelove* when he offered as the Strategic Air Command's motto "Peace Is Our Profession." Keeping, or even *making,* peace seemed to be a preoccupation of the Cold War mind-set, and for good reason. Just as a mind that is not at peace is chaotic and irrational, subject to all sorts of unpredictable flights of emotion, so, too, was a world not at peace a daunting, terrifying prospect. The rational, even, *scientific* approach was best suited to the maintenance of a peaceful mind and, by extension, a peaceful world.

This comparison may seem strained, but a predominate motivating factor in nearly all Cold War policy and strategy was the idea that we were facing a global situation whose outcome carried with it unprecedented consequences. It is little wonder that the people who came to Dr. Popenoe were "so confused by worry that they [could not] think clearly" when the fate of the world was on their conscience. One's family was, of course, the most reasonable and immediate cause for concern, and if Massachusetts Mutual could cash in on a bit of this anxiety, keeping the economic cycle tirelessly moving, this only spoke to the innovativeness of the successful capitalistic entrepreneur. Dr. Popenoe's endorsement was, like the attempt to educate the public using eminent scientists in the *Scientific American* features, a method of appeasement, a furthering of the solubility ethos, a suggestion that rational, reductionistic thinking would provide the coping tools needed to see U.S. citizens through the paralyzing crisis that was the Cold War. The idea of peace was metaphorically offered as a way to think about one's own mental health; it was a symbolic goal that represented both the hopes of the country and the needs of the individual, and, ultimately, it resulted in a nation of people who could operate with control and smooth performance.

The Metaphorical Cycle

Scientific/technological metaphors manifested themselves in nearly every sector of American society during the early Cold War, but the metaphors used during the Cold War in general pointed to the need to enlist people into and maintain the scientific program. The very definition of the circumstances outlined by Winston Churchill early in the conflict was expressed in metaphorical terms, dictating for the American public a dominant way of thinking about their situation, its implications, and the measures that must be taken to address it. The Iron Curtain metaphor, while not in itself a scientifically oriented device, led to policies that were very much scientific and technological at their core. In fact, the military nature of the metaphor assumed a technological solution to the conditions that faced the country, and a precedent had already been set regarding our approach to threatening adversaries in the form of the atomic bomb. But the bomb was only one rather specific manifestation of the technological steps needed to create a strong and formidable nation; strength in technology implied strength in economy, which implied strength in agriculture and industry and a population committed to the program of scientific/technological development, innovation, and expansion.

The result was a rhetorical network of symbols that were often metaphorically based, and familiar religious references were frequently found in these metaphors. I have provided just a few examples as a way to point to the metaphorical nature of popular communication, but even the casual perusal of popular journalistic texts of the time reveals that certain metaphors dominated the thinking of the American public, at least as it was reflected by such documents, and in a way that spread out over all fronts of American society. The apparent purpose was to create a unified ideological front that was interdependent, emphasizing that strength in the factory or on the farm was tantamount to strength as a nation. The logic is inescapable on many levels; it makes sense, for example, that efficient business produces the economic stability and industrial outpouring necessary for a well-established military presence and that technology was a key factor in this equation, and the social effect this had cannot be overstated. By creating a sense of purpose, order, efficiency, and security, the media were able to address the main concerns of the public in a way that helped foster unity and a common goal.

The difficulty is determining whether these measures had the desired rhetorical effect. Certainly one could not easily avoid the patriotic program being endorsed by the distributors of information, and it is perhaps too convenient to label the activities of the media as mere propaganda. By my estimation, the results were more subtle than this. The sense of purpose generated by the metaphors of the Cold War suited a need—a need for reassurance and the maintenance of social integrity, an understanding that being part of the program meant the safeguarding of a way of life that, while overstated and caricatured on many levels, hardly required great waves of propaganda to enlist public support. Americans were conditioned to accept the patriotic assumptions implicit in a technological/scientific program because they had seen it in action—working toward growth, improving the standard of living, and solving problems that otherwise seemed insurmountable. The ubiquity of scientific metaphors was only one way such tropes were used in support of science. Scientific concepts and theories were necessarily reduced to metaphors because this was an accessible way to deliver scientific ideas. Such metaphors are so dominant in major scientific theories that we have difficulty conceiving of them in any other way. We think, for example, of evolution and natural selection in terms of a ladder because this is how it has been explained to us, and this is, therefore, how we visualize the idea. We think of evolution as a progress from a lower form of life to a higher one—a value judgment superimposed on an innately neutral (and indifferently natural) process. A metaphor like this is actually inaccurate; evolution is not a series of improvements in the sense that we normally think of the word. A more accurate metaphor, as Stephen Jay Gould points out, is provided in a branching image, where many organisms, in an attempt (a word that is in itself inaccurate because it implies an active will to evolve) to increase the likelihood of survival, diverge from the main, often producing variations that are in fact not improvements at all but failed evolutionary experiments, ending like the tip of a branch that may grow but will eventually reach its maximum length. Such a metaphor also accounts for the remarkable variety of life on earth. Because creatures are continuously changing, albeit very slowly, they are constantly mutating into different forms and going in different directions—some successful and some not. Improvement and progress, then, are erroneous notions imbedded in the ladder metaphor since organisms do not improve; they merely make adjustments that may or may not be effective. But it is because we have been culturally conditioned to embrace notions of progress that the ladder metaphor thrives—a kind of rhetorical evolution that, ironically, parallels our understanding of natural selection .

So metaphors are both enlightening and limiting; they illuminate in accessible terms ideas that otherwise would require considerable training and experience to understand. The Iron Curtain metaphor, likewise, provided only one possible perspective on a much more complex situation, but it dominated American thinking about the Cold War because it was delivered by a diplomatic authority who was trusted, admired, and respected. The command of the metaphor resulted in a preordained strategy that involved science as much as it presupposed patriotic acceptance. In fact, the scientific approach was implicit in the patriotic policy of containment; a nation could not stave off the aggressive, self-interested expansion of an adversary like Russia without the technological tools to do so. Most of these tools were military—planes, bombs, ships, submarines, etc.— but they were much more encompassing than that. The very standard of living that Americans were so anxious to protect was a technological way of life and it was invested in a scientific program of progress. The cycle was apparent: in order to retain the comforts of modern living—the public and private machines that made life ostensibly easier—and the nation had to be committed to the technological and economic production of these same machines. It is little wonder that American thinking was steeped in the machine metaphor; they were as much a part of the social mechanism that made the machines go as were the machines contributing to the way of life that Americans had been enjoying for at least two decades. Industry was the "bulwark of freedom" because it contributed so much to the lifestyle of Americans, not just on a physical level of production but in the very pattern of thinking that Americans took as absolute.

The metaphorical cycle, then, is one that reflected a common orientation and created a level of understanding that was imperative to the nation's economical, social, and international health. These tropes at once buttressed preconceived ideologies, drawing on the myths and assumptions that Americans took as given, and they supplied an additional level of awareness for the scientific and ethical questions that preoccupied the public consciousness. At the same time, they functioned as a lens that limited the social perspective by magnifying certain desirable features of the Cold War condition or patriotic idea at the expense of other, equally valid, perceptions. The entire approach to the postwar world created a mood that was dictated by the visual image of the Iron Curtain and the policy of containment. The machine metaphor was used to emphasize the need for technological evolution and to reinforce the notion that efficiency and productivity would be the answer to our global and domestic problems, and it fulfilled a need to create a unified social front where all Americans were working toward a common end and under a common purpose.

We do not often consider the truly powerful impact that language can have on our social outlook, but a close examination of key metaphors helps underscore the unifying effect language can have on society. In a country where action is valued and mere rhetoric is suspect, it may come as a shock to realize that these activities are not by any means mutually exclusive. Language is, as Kenneth Burke so astutely pointed out, symbolic action; no meaningful, deliberative activity takes place without first establishing a communicative (that is, rhetorical) cooperation. The beauty of metaphors is that they are an efficient means of capturing popular sentiment and expressing dominant ideologies, and from a rhetorical perspective, they refine a cooperation that is already established but sometimes dormant, sometimes physically active. But metaphors do more than trumpet already extant orientations; they provide a degree of accessibility to ideas that might otherwise require considerable education and training. The metaphors of the Cold War reflected the public mind-set, but they also expanded and limited it. The complexity of the situation and the needs of the country required the use of communicative devices that could pull people together actively and symbolically. Metaphors, far from being just the handy poetic tools our grammar school teachers once claimed they were, helped ideologically integrate a country, maintain and develop a program of policy, and guide our understanding of issues and ideas in a world that at once provoked confusion and offered promise.

Conclusion
Cold War Leftovers

While the politicians and the media like to claim these days that the Cold War is over (the Soviet Union as a military and political threat is ostensibly dissolved, and the credit for this situation has been taken by a number of individuals and agencies), the attitudes, assumptions, and ideologies that drove the Cold War are impressed so deeply on us as a society that the by-products of the Cold War mentality do not fade as easily. We continue to think in terms of technological and scientific progress, especially as it affects U.S. status as a global superpower. Consider, for example, George W. Bush's policy calling for increased education in mathematics, the sciences, and technology as a way to keep us ahead of the competition in international economics (learning that is evaluated, incidentally, through the use of quantifiable standardized tests, revealing an assumption that knowledge is cumulative and instantly reproducible, just as an experiment is predictable and certain). In one respect, this is a prudent and sensible move; in another, it reflects an attitude that has dominated American thinking since the end of World War II, thinking that has, at least once in the last forty years, nearly brought us to the brink of civil destruction.

While we can speculate about whether the development of science as a pervasive linguistic umbrella was a deliberate act of authoritative enabling on the part of scientists, this seems to me irrelevant. It is clear that it has become so, regardless of the intentions of scientists and laity alike. Examples of science used as a rhetorical appeal to authority are bountiful. In advertising, quasi-scientific charts, tables, statistics, endorsers, comparisons, and rebuttals are used to sell Americans thousands of products that are neither needed nor included under a strict rubric accountable to scientific inspection. In law, attorneys employ scientific experts to argue for or against certain scientific facts, and often these experts violently disagree. In our personal lives, we enlist the services of counselors, therapists, doctors, practitioners, specialists, and technicians to fix whatever we may have in body, mind, or property that needs fixing. We also have a seemingly inexhaustible fascination with gadgetry, always wanting to stay one step ahead technologically by purchasing the toy with the greatest number of buttons and features. And that science and capitalism enjoy such a reciprocal relationship speaks to the way the scientific perspective can so easily flourish in our society.

The emphasis on science and technology is still very strong, and to a great extent, it should be. Ours is a technological society and, as such, demands knowledge of the sciences to function properly. On the other hand, if Burke is correct in saying that scientistic thinking preoccupies us with the motion of objects rather than the action of subjects, then we are preordained to commit the same mistakes that launched us into the Cold War in 1945. Rather than emphasizing one mode of thinking over another, a coalescence of the sciences and humanities—a system that checks and balances itself to include the practical as well as the aesthetic and the ethical—might be beneficial to the well-being of our country and society. It is erroneous to assume that humanistic education is without its practical merits. Studies in philosophy, history, literature, rhetoric, writing, and the fine arts do more than appease territorial faculty at liberal arts institutions; they provide an intellectual and critical basis for decision making and problem solving, helping students understand themselves and the world around them so they can make informed, intelligent, and innovative choices. Unfortunately, too many people see humanistic endeavors as irrelevant or, worse, a waste of time and resources. General education programs that attempt to expose students to the value of a variety of knowledge systems are being continually reformed in an attempt to cut the fat; the fat usually consisting of humanities courses—music, literature, and art—that many students (and even administrators) see as a scheme to soak up student tuition for no apparent educational reason or immediately applicable financial return. Recently, for example, Loyola University of Chicago disbanded its classics department because there was no longer a demand for the services they provided (the irony being, of course, that if a Jesuit university has no need for the classics, who does?). Foreign language departments across the nation have been under fire from university administrations because the learning they have to offer is not considered pertinent to modern living—despite the fact that the United States has a greater number of immigrants and bilingual speakers than ever. In universities, the business model rules, and the bottom line is whether an institution can sell credits or increase enrollment by offering students only the utilitarian foundation they need to function technologically and economically. Education has given way to training and skills, and the great bulk of the more practical (post)modern curriculum consists of courses in science and technology (check any university's enrollment figures on which majors students most frequently declare; you'll find that computer science majors have doubled in the last decade. The favored status of science and technology that was a mainstay of Cold War thinking is still very much with us, and the conservative push to return to this apparently commonsensical bygone era is in the forefront of many peoples' minds.

The writing of this book has been, from the beginning, an observational exercise, intended primarily to show that we harbor certain cultural/philosophical assumptions about science that do not fully acknowledge the extent of its rhetorical contribution to American thinking. Its purpose from a scholarly standpoint has been unapologetically interdisciplinary, because it is only by employing the methodologies of history, philosophy, religion, rhetoric, cultural studies, and literary criticism that we can get anything like a comprehensive picture of the American cultural landscape during the early Cold War. And even after drawing upon the strengths of these fields, the result is still only a snapshot of the whole panorama of the American experience. But it is science that was the primary focus in this project, for it is science that offers the greatest promise for our future. Because it is so deeply embedded in our way of thinking, science is also the most susceptible to abuse and complacency. Science, apart from the recent (ostensibly postmodern) trend of describing it as a socially constructed entity, is also a language system designed to impart very specific kinds of information for very specific purposes, operating under very specific and limiting assumptions about the way the world works. More important, science is so much a part of our culture that it is frequently (and often erroneously) called upon by the nonscientist to elicit effects that are essentially—and equally—nonscientific. The early Cold War was a major source for this precedent because science was viewed at this time as the national savior, the approach that would conquer any and all problems and barriers. Scientists and statesmen alike campaigned to support not only scientific advancement, but also scientific education on a broad but limited scale. Science enjoyed its status because it had governmental, ideological, and public support—support that it had earned but that was also manipulated for purposes that had little to do with the pure, ideal pursuit of scientific knowledge. Agendas existed that needed science as an advocate and as a method, and few people seriously questioned science's role in these adjacent arenas of social activity.

Scientists themselves were in an unusual, and by no means uniform, position during this time, and while I have been forced at times to make some rather broad generalizations, I hope I have also demonstrated that while the role of the scientist became fairly well defined during the postwar years, the way that scientists viewed themselves often differed significantly. Not all scientists were in agreement with national policy or the development of weapons of mass destruction, but it is interesting to note that only the most eminent scientists were supplied with a political voice to debate the issue. The majority of scientists were working for corporations that had governmental contracts to develop specific kinds of technology, whether it be thermonuclear weapons, experimental aircraft, or new materials and procedures to be used for space exploration. The economic prosperity that dovetailed with technological advancement meant that the scientific program was an integral part of society's cultural makeup. The language of science was therefore part of the national identity, even though it was a language that had been simplified and condensed for mass consumption.

What does this mean for science as a cultural entity, one that is used rhetorically as a means to a social end outside the realm of uncorrupted scientific discovery? It means we might benefit from viewing science as more than a neutral analytical tool designed exclusively for data collection and interpretation. Even the most conservative scientist would concede that it is much more than this. A synthesis has been reached in our scholarly approach to the effects of scientific discourse: people have swung from a naive view of science as an objective methodology to a view of science as purely a social construct—no more real or unreal than any other social construct—to arrive at a more moderate position of understanding that science is in fact much more complicated than either of these extremes. It is a method and a philosophy, an ideal and a reality, a practice and an abstract system. It touches the social structure because it does not operate in the laboratory alone; it would be of little use if it did. It is part of the lexicon, and as such, guides individuals' thinking, just as all language contributes to a perspective on the environment. It is important to remember this if only to understand the cultural impact science has had on the United States and to avoid the pitfalls and anthropocentric hubris that can accompany a feeling that any one knowledge system holds the key to understanding ourselves and the universe. Put in its proper perspective—understanding that science is a powerful, language-based tool that can and has brought about much good but has also brought about much pain—science allows humans to know themselves and their role on this planet better.

I have argued that science is an ideology and a language as much as it is a method, using a specific development in American history as a representative focal point to elucidate this claim. The difficulty in achieving this has been in the sheer complexity of the ideological structure that governs human activity, drawing on the elements of one paradigm while concurrently using the language of another and vice versa. The distinction between science and religion is one that has been cultivated historically but in fact does not exist to the extent generally believed.

The Cold War is only one possible period of study to begin a fuller understanding of how science affects the public on levels that go beyond the rational or factual to examine the more deeply rooted cultural framework that makes U.S. society one that responds loyally to claims and ideas that incorporate a scientific discourse. Science is a very important tool for examining the world and our place in it, but when we are blinded to its limitations, seeing it as an infallible system (despite the fact that it was created by fallible beings), we make a dangerous mistake—one that makes the practitioners of science corruptible and arrogant and the recipients of scientific advances complacent and vulnerable. Practicing the scrutiny that scientists advocate should apply to all learning and knowledge, and sometimes even to the language of science itself. We might ask ourselves, for example, if the scientific method of careful observation and data collection is adequate to tell us anything definite about what we want to know about the human mind or human society. Will the scientific method suffice, for example, to provide a full, rich interpretation of a literary text or work of art? The way that the individual human mind operates or how social groups function? Are there other methods and processes that might supply added understanding or perspectives that can supplement, enhance, or even replace the scientific method? Will too much reliance on scientific data and scientific conclusions make us a less tolerant society? Will this same reliance paralyze us if we covet scientific knowledge exclusively?

As a species, human beings crave order and coherence. We have well-worn standards of what is acceptable in this respect and what is not. Scientific discourse, with all its heritage, predictive precision, and rigorous criteria, is still a discourse and an ideology, subject to the barriers and pitfalls of human error and human egoism that plague other ordering systems. Its philosophical dictates are sound; when used in tandem with other forms of human knowledge equally sound, it can be the best tool we have for achieving understanding of the world and ourselves. If we delude ourselves into thinking that it is always strictly neutral and objective—that it is not, in short, governed by prejudices similar to those of all other language systems—then it is practiced without the standards of judgment required to make it a fully human, accountable enterprise. As the Cold War has shown, the advances of science are not always for the better; when we look to science as the only worthwhile guide in a labyrinth of complex domestic, social, and international politics, we create an illusion that overlooks the linguistic magnitude of human activity that touches *all* ordering systems and provides the basis for *all* human motivations.

Notes

Chapter 1

1. *Language* is used here in its broadest context. While language is often thought of as it appears in formally written or spoken form, these are only the least of its representations. Language, in the broad sense, is any act or recognition of representation that communicates symbolically. The raising of one eyebrow (in American culture, at least) is an act of language in that it communicates a basic idea symbolically—that is, that one is puzzled or doubtful. Colors are linguistic in particular contexts, as when red appears in octagonal form on top of a pole at the corner of an intersection. Architecture or interior design is linguistic in that it directs people into certain spaces in certain ways, thus communicating a preference by its designers for people to be channeled into specific areas for specific purposes. Clothing is another example. A defense lawyer who argues that his client, who is on trial for rape, was provoked by the victim's clothing suggests that the victim *communicated* the desire to have sex (we see in this example how language can be easily abused to reach an erroneous conclusion). While I will not necessarily concentrate on these more hidden forms of language, it is necessary to think of language in its more sweeping applications, lest I be accused of using the term in too liberal a manner.

2. In *A Grammar of Motives,* Burke describes science as having five major aspects: (1) a high development of technological specialization; (2) an involvement with rationale of money (accountancy); (3) a progressive departure from natural conditions, usually saluted in the name of naturalism; (4) a reduction of scenic circumference to empirical limits; and (5) a stress upon the problem of knowledge as the point of departure for philosophic speculation (515). This may prove, however, to be a bit too constraining. While accurate in the rhetorical and

linguistic sense, this description is specialized rather than generalized. Let us say, then, that science is a philosophical and methodological ideology that, apart from its ability to represent the natural world in empirical terms, carries with it the rhetorical features of an institution since it limits, delimits, and defines specific linguistic parameters of accepted fact and measurable knowledge in a way that controls how new knowledge might be constructed and imparted.

3. We will see how this operates more explicitly in chapter 3 with such texts as Dwight Eisenhower's "Atoms for Peace" speech and Winston Churchill's "Iron Curtain" address.

4. I use Gramsci's idea of hegemony as a totalizing ideology, one that permeates all levels of social consciousness. It is, in the words of Raymond Williams, an ideology "which is lived at such a depth, which saturates society to such an extent, and ... even constitutes the substance and limit of common sense for most people under its sway, that it corresponds to the reality of social experience very much more clearly than any notions derived from the formula of base and superstructure" (382).

5. Burke notes that "'substance' is an abstruse philosophic term, beset by a long history of quandaries and puzzlements. It names so paradoxical a function in men's systematic terminologies, that thinkers finally tried to abolish it. ... They abolished the *term,* but it is doubtful whether they can abolish the *function* of that term, or even whether they should *want* to" (*Rhetoric of Motives* 21).

6. I should also note that when certain ideological motifs become dominant, metaphors frequently emerge to describe all sorts of things outside the sphere of the ideology itself. In other words, if efficiency becomes a dominant ideological quality, then comparisons to machinery are often made to reflect this ideal. An example would be to refer to one's football team as a well-oiled machine, suggesting that each part has a specific function that works in concert with all the other parts to create an efficient human mechanism that always functions exactly as it is supposed to, without waste in energy or movement. Such metaphors only have meaning in a society that identifies with the ideology that the metaphor is meant to represent, in this case, the positive qualities of a winning team. Interestingly, this particular metaphor is often used in military circles to describe a platoon, squad, or other subgroup of an army.

7. When I provided this example to my wife, she regarded it as a possible Freudian slip, indicating that my real motive for using the example was that I might in fact subconsciously want her dead. She was joking, of course, but her inability to recognize that she was employing the very psychoanalytic orientation I was critiquing indicates how accustomed we are to invoking this type of analysis. Psychoanalysis is so socially ingrained that we routinely call upon its language to describe behavioral motives.

8. It should be noted that literary critics sometimes make the same mistake by assuming that the language they have developed to describe literary operations is somehow beyond the obstacles of the language they are studying. Taken

to its furthest extreme, this problem, if recognized, can turn into something like deconstruction, showing not only that language can never shake off adjacent associations, but also that all language is in fact meaningless. The problem with this is obvious: it is not that language ultimately has *no* meaning, but that it ultimately has *too much* meaning.

Chapter 3

1. The author, Ralph E. Lapp, goes so far as to say that the detonation of three dozen 500-kiloton hydrogen bombs "would not produce a serious health hazard from prolonged radioactivity" (14). While one explanation of this cavalier statement is that Lapp must be speaking in relative terms (suggesting that damage by blast and fire would be far more destructive than the effects of radiation), it is equally possible that the long-term effects of nuclear radiation were not yet fully studied or understood. It is plausible to speculate that if these effects *were* in fact understood, it was deemed prudent not to reveal them to the public at large lest scientific authorities panic the nation's population even further. Moreover, the editors of *Scientific American* either did not believe that radiation was that harmful, knew it was harmful but wanted to disarm public anxiety by calling more reactionary scientists "prophets of doom" (12), or felt that the priority for civil defense lay with protection against blast and fire. In any event, there have been some clear rhetorical editing decisions in Lapp's statement, who probably realizes that the subtler details of the dangers of radiation poisoning might be lost on the general public.

Chapter 4

1. Another interesting irony here is that Eisenhower's main purpose behind the interstate system was not free and easy travel for the American citizen. While this was a convenient and happy economic side effect, the real intent was to create a massive evacuation and National Defense network. While looking at a road atlas, note all the federal interstate superhighways built during the Eisenhower administration and the pattern in which they are laid out: they function roughly as a hub and web configuration, originating at the larger urban centers and spinning out efficiently from their center. The long main fingers of the roads reach to all coastal corners and seaboards. In the event of a military emergency or even an invasion, these roads were to be used for two things: civil evacuation away from the main urban centers that would have been targets during a nuclear war, and tanks, trucks, troop carriers, and other mechanized military equipment that could quickly and effectively reach the coastline to defend American soil. Even today, military convoys use interstates for the mobilization

of equipment, troops, and ordnance since most state, county, and local highways are too narrow and convoluted to make efficient travel routes. This again demonstrates the clear and inescapable connection between domestic prosperity, economic expansion, and military readiness.

2. Corporate science would be an interesting study in its own right. Current examples of how science can be manipulated and scientists bought off to create desirable findings are plentiful, as in the case of the tobacco company scientists who attempt to counter the unfavorable findings of the surgeon general for legal and economic reasons. Other examples include oil company scientists who assert that ethanol is worse for the environment than petroleum-based gasoline, despite strong evidence that grain alcohol burns much cleaner than oil-based fuels, being free of many of the organic impurities found in traditional gasoline. The environment, in fact, is a virtual debaters' forum for competing scientific viewpoints, each vying for funds and recognition and jockeying for position in the public and governmental spotlight. These are obvious attempts to fight science with science, and it is clear, for anyone who may doubt the rhetorical nature of scientific activity, that facts and data do *not* speak for themselves, but require a vocal human agent to interpret, coordinate, and supplement findings for a desired social, economic, or legal end. In short, when there is a profit base at stake, scientists can be employed (and prostituted) in much the same way that any other expert can, and even when scientists maintain their integrity and report just the facts, they may find themselves out of a job or have their results repressed in an effort to create the most affirmative image of the corporate identity.

Bibliography

Note: World Wide Web addresses were current as of the publication of this book.

Adams, Gordon. *The Politics of Defense Contracting.* New Brunswick, N.J.: Transaction Books, 1982.

Adams, John Charles. Review of *A Rhetoric of Science: Inventing Scientific Discourse,* by Lawrence J. Prelli. *Rhetoric Review* 9.2 (spring 1991): 369-72.

Adams, Sherman. *Firsthand Report: The Story of the Eisenhower Administration.* New York: Popular Library, 1961.

"AEC Report." *Scientific American* 189.3 (September 1952): 70-72.

Anisfield, Nancy, ed. *The Nightmare Considered: Critical Essays on Nuclear War Literature.* Bowling Green, Ohio: Bowling Green State University Popular Press, 1991.

"Antivivisectionists in Retreat." *Scientific American* 189.1 (July 1953): 47-48.

Aronson, James. *The Press and the Cold War.* Boston: Beacon Press, 1970.

"Atomic Radiation Preserves Food." *Life,* March 16, 1953, 47-52.

"Atomics, Unlimited." *Scientific American* 186.6 (June 1952): 40.

Axelsson, Arne. *Restrained Response: American Novels of the Cold War and Korea, 1945-1962.* New York: Greenwood Press, 1990.

Bacher, Robert F. "The Hydrogen Bomb: III." *Scientific American* 182.5 (May 1950): 11-15.

Bendix Aviation Corporation. Advertisement. *Scientific American* 186.5 (May 1952): 38.

Bennett, Michael Alan. "The Theme of Responsibility in Miller's *A Canticle for Leibowitz.*" *English Journal* 59 (April 1970): 484-89.

Berger, Peter L., and Thomas Luckmann. *The Social Construction of Reality.* New York: Anchor Books, 1967.

Bethe, Hans A. "The Hydrogen Bomb: II." *Scientific American* 182.4 (April 1950): 18–23.

"Biggest Sphere for the Atomic Sub Engine." *Life,* December 15, 1952, 62–75.

Blankenship, Jane, and Janette Kenner Muir. "On Imaging the Future: The Secular Search for 'Piety.'" *Communication Quarterly* 35 (winter 1987): 1–12.

Boeing Airplane Company. Advertisement. *Scientific American* 186.2 (February 1952): 39.

Bogart, Leo. *Cool Words, Cold War: A New Look at USIA's Premises for Propaganda.* Washington, D.C.: American University Press, 1995.

———. *Premises for Propaganda: The United States Information Agency's Operating Assumptions in the Cold War.* New York: Free Press, 1976.

Born, Daniel. "Character as Perception: Science Fiction and the Christian Man of Faith." *Extrapolation* 24 (1983): 251–71.

Boyer, Paul. *By the Bomb's Early Light: American Thought and Culture at the Dawn of the Atomic Age.* New York: Pantheon, 1985.

Brians, Paul. *Nuclear Holocausts: Atomic War in Fiction, 1895–1984.* Kent, Ohio: Kent State University Press, 1987.

Brockriede, Wayne, and Robert L. Scott. *Moments in the Rhetoric of the Cold War.* New York: Random House, 1970.

Bromo-Seltzer. Advertisement. *Life,* December 15, 1952, 14.

Brown, Richard Harvey. "Reason as Rhetorical: On Relations among Epistemology, Discourse, and Practice." In *The Rhetoric of the Human Sciences: Language and Argument in Scholarship and Public Affairs,* ed. John S. Nelson, Allan Megill, and Donald N. McCloskey. Madison: University of Wisconsin Press, 1987.

Bruns, Gerald L. "On the Weakness of Language in the Human Sciences." In *The Rhetoric of the Human Sciences: Language and Argument in Scholarship and Public Affairs,* ed. John S. Nelson, Allan Megill, and Donald N. McCloskey. Madison: University of Wisconsin Press, 1987.

Bundy, McGeorge. *Danger and Survival: Choices about the Bomb in the First Fifty Years.* New York: Vintage Books, 1988.

Burhop, E. H. S. "Scientists and Public Affairs." In *Society and Science,* ed. Maurice Goldsmith and Alan Macay. New York: Simon and Schuster, 1964.

Burke, Kenneth. *Attitudes toward History.* 3d ed. Berkeley: University of California Press, 1984.

———. *Counter-Statement.* 2d ed. Berkeley: University of California Press, 1968.

———. *A Grammar of Motives.* Berkeley: University of California Press, 1969.

———. *Language as Symbolic Action: Essays on Life, Literature, and Method.* Berkeley: University of California Press, 1966.

———. *Permanence and Change: An Anatomy of Purpose.* 3d ed. Berkeley: University of California Press, 1984.

———. *The Philosophy of Literary Form.* 3d ed. Berkeley: University of California Press, 1973.

————. *A Rhetoric of Motives*. Berkeley: University of California Press, 1969.

————. *The Rhetoric of Religion: Studies in Logology*. Berkeley: University of California Press, 1970.

Brush Electronics Company. Advertisement. *Scientific American* 189.2 (August 1953): 4.

Campbell, Angus, Gerald Gurin, and Warren E. Miller. "Television and the Election." *Scientific American* 188.5 (May 1953): 46-48.

Campbell, John Angus. "Charles Darwin: Rhetorician of Science." In *The Rhetoric of the Human Sciences,* ed. John S. Nelson, Allan Megill, and Donald N. McCloskey, 69-86. Madison: University of Wisconsin Press, 1987.

Carboloy Metals. Advertisement. *Scientific American* 186.5 (May 1952): 44.

Chilton, Paul A. *Security Metaphors: Cold War Discourse from Containment to Common House*. New York: Peter Lang, 1995.

Clark, Suzanne. *Cold Warriors: Manliness on Trial in the Rhetoric of the West*. Carbondale: Southern Illinois University Press, 2000.

Clarke, Arthur C. "Will a Hungry World Raise Whales for Food?" *Popular Science* 177.5 (November 1960): 74-82.

Cohen, Carol, ed. *Benet's Reader's Encyclopedia*. 3d ed. New York: Harper and Row, 1987.

Connolly, William E. *The Terms of Political Discourse*. 2d ed. Princeton, N.J.: Princeton University Press, 1983.

Cragen, John F. "The Origins and Nature of the Cold War Rhetorical Vision, 1946-1972: A Partial History." In *Applied Communication Research: A Dramatistic Approach,* ed. John F. Cragen and Donald C. Shields. Prospect Heights, Ill.: Waveland Press, 1981.

Crusius, Timothy W. *Kenneth Burke and the Conversation after Philosophy*. Carbondale: Southern Illinois University Press, 1999.

Curtin, Michael. *Redeeming the Wasteland: Television Documentary and Cold War Politics*. Piscataway, N.J.: Rutgers University Press, 1995.

Dampier, William C., and Margaret Dampier, eds. *Readings in the Literature of Science*. New York: Harper and Brothers, 1959.

Darrow, Karl K. "The Quantum Theory." *Scientific American* 186.3 (March 1952): 47-54.

Darwin, Charles. Excepts from *The Descent of Man*. *Readings in the Literature of Science*. Ed. William C. Dampier and Margaret Dampier. New York: Harper and Brothers, 1959.

Dyck, Joachim. "Rhetoric and Psychoanalysis." *Rhetoric Society Quarterly* 19.2 (spring 1989): 95-104.

Dyson, Freeman. *Weapons and Hope*. New York: Harper and Row Publishers, 1984.

"The Earth Is a Hothouse." *Scientific American* 189.1 (July 1953): 44-45.

Eaves, Lindon. "Spirit, Method, and Content in Science and Religion: The Theological Perspectives of a Geneticist." *Zygon* 24.2 (June 1, 1989): 185.

Ellul, Jacques. *Propaganda: The Formation of Men's Attitudes.* New York: Alfred A. Knopf, 1968.

Fahnstock, Jeanne. "Arguing in Different Forums: The Bering-Crossover Controversy." In *Landmark Essays on Rhetoric of Science,* ed. Randy Allen Harris. Mahwah, N.J.: Hermagoras Press, 1997.

Fenwal Electric Temperature Control and Detection Devices. Advertisement. *Scientific American* 188.2 (February 1953): 43.

Feyerabend, Paul. *Against Method: Outline of an Anarchistic Theory of Knowledge.* Atlantic Highlands, N.J.: Humanities Press, 1975.

———. *Science in a Free Society.* London: NLB, 1978.

———. *Three Dialogues on Knowledge.* Cambridge, Mass: Blackwell Publishers, 1991.

Feyerabend, Paul, and Grover Maxwell, eds. *Mind, Matter, and Method: Essays in Philosophy and Science in Honor of Herbert Feigl.* Minneapolis: University of Minnesota Press, 1966.

Fish, Stanley. "Professor Sokal's Bad Joke." *New York Times,* May 21, 1996, A23. Available online at http://www.drizzle.com/˜jwalsh/sokal/articles/fish-oped.html.

Forbes, R. J., and E. J. Dijksterhuis. *A History of Science and Technology.* Vol. 2, *Nature Obeyed and Conquered: The Eighteenth and Nineteenth Centuries.* Baltimore: Penguin Books, 1963.

Foss, Karen A., and Stephen W. Littlejohn. "The Day After: Rhetorical Vision in an Ironic Frame." *Critical Studies in Mass Communication* 3 (1986): 317-36.

Foucault, Michel. *Discipline and Punish: The Birth of the Prison.* Trans. Alan Sheridan. New York: Vintage Books, 1979.

———. *The Order of Things: An Archaeology of the Human Sciences.* New York: Vintage Books, 1973.

Freedman, Lawrence. *The Evolution of Nuclear Strategy.* London: Macmillan, 1989.

Fuller, Steve. "'Rhetoric of Science': A Doubly Vexed Expression." *Southern Communication Journal* 58.4 (summer 1993): 306-11.

Gaonkar, Dilip Parameshwar. "The Idea of Rhetoric in the Rhetoric of Science." *Southern Communication Journal* 58.4 (summer 1993): 258-95.

General Cable Corporation. Advertisement. *Time,* January 7, 1952, 61-62.

General Electric Corporation. Advertisement. *Life,* March 16, 1953, 25.

"Gift on a 74th Birthday." *Life,* March 30, 1953, 112-15.

Gilpin, Robert. *American Scientists and Nuclear Weapons Policy.* Princeton, N.J.: Princeton University Press, 1962.

"Government Research in Colleges." *Scientific American* 188.5 (May 1953): 53-54.

Greene, John C. "The Kuhnian Paradigm and the Darwinian Revolution in Natural History." In *Perspectives in the History of Science and Technology,* ed. Duane H. D. Roller, 3-25. Norman: University of Oklahoma Press, 1971.

Gregory, Donna, ed. *The Nuclear Predicament: A Sourcebook.* New York: St. Martin's Press, 1986.

Griffin, Russell M. "A Medievalism in *A Canticle for Leibowitz*." *Extrapolation* 14 (1973): 112–25.

Gross, Alan G. "Does Rhetoric of Science Matter? The Case of the Floppy-Eared Rabbits." *College English* 53.8 (December 1, 1991): 933–41.

———. "On the Shoulders of Giants: Seventeenth-Century Optics as an Argument Field." In *Landmark Essays on Rhetoric of Science*, ed. Randy Allen Harris. Mahwah, N.J.: Hermagoras Press, 1997.

———. "Rhetoric of Science Is Epistemic Rhetoric." *Quarterly Journal of Speech* 76 (1990): 304–6.

———. "Rhetoric of Science without Constraints." *Rhetorica* 9.4 (fall 1991): 283–99.

———. "What If We're Not Producing Knowledge? Critical Reflections on the Rhetorical Criticism of Science." *Southern Communication Journal* 58.4 (summer 1993): 301–5.

Grove, J. W. *In Defense of Science: Science, Technology, and Politics in Modern Society*. Toronto: University of Toronto Press, 1989.

Halloran, S. Michael. "The Birth of Molecular Biology: An Essay in the Rhetorical Criticism of Science." In *Landmark Essays on Rhetoric of Science*, ed. Randy Allen Harris. Mahwah, N.J.: Hermagoras Press, 1997.

Harvey, Frank. "War Room in the Sky." *Popular Science* 177.5 (November 1960): 94–218.

Hendershot, Cyndy. *Paranoia, the Bomb, and 1950s Science Fiction Films*. Bowling Green: Bowling Green State University Popular Press, 1999.

Herbert, Gary B. "The Hegelian 'Bad Infinite' in Walter M. Miller's *A Canticle for Leibowitz*." *Extrapolation* 31.2 (1990): 160–69.

Hicks, James E. "A Selective Annotated Bibliography of *A Canticle for Leibowitz*." *Extrapolation* 31.3 (1990): 216–28.

Hilgartner, Stephen, Richard C. Bell, and Rory O'Connor. *Nukespeak: Nuclear Language, Visions, and Mindset*. San Francisco: Sierra Club Books, 1982.

Hinds, Lynn Boyd, and Theodore Otto Windt Jr. *The Cold War as Rhetoric*. Westport, Conn.: Praeger, 1991.

Hinkins, James W. "The Rhetoric of 'Unconditional Surrender' and the Decision to Drop the Atomic Bomb." *Quarterly Journal of Speech* 69 (1983): 379–400.

Honner, John. "Science vs. Religion." *Commonweal* 121.16 (September 23, 1994): 14.

Horvat, Robert E. "The Science Education for Public Understanding Program: What's New with SEPUP." *Bulletin of Science, Technology, and Society* 13.4 (1993): 208.

Horwich, Paul, ed. *World Changes: Thomas Kuhn and the Nature of Science*. Cambridge, Mass.: MIT Press, 1993.

Hunt, Everett. "The Rhetorical Mood of World War II." *Quarterly Journal of Speech* 29 (1943): 1–5.

Index of Advertisers. *Scientific American* 189.1 (July 1953): 99.

Itzkoff, Seymour W. *Ernst Cassirer: Scientific Knowledge and the Concept of Man*. 2d ed. Notre Dame: University of Notre Dame Press, 1997.

Ivie, Robert L. "Metaphor and Rhetorical Invention of Cold War 'Idealists.'" *Communication Monographs* 54 (1987): 165–82.

Johnston, George Sim. *The Galileo Affair*. Princeton: Scepter Press. August 17, 1996. Available online at http://www.catholic.net/rcc/Periodicals/Issues/Galileo.html.

Jones, Greta. *Science, Politics, and the Cold War*. London: Routledge, 1988.

Jones, W. T. *A History of Western Philosophy: Hobbes to Hume*. San Diego: Harcourt Brace Jovanovich, 1969.

Kerr, Thomas J. *Civil Defense in the US: Band-Aid for a Holocaust?* Boulder, Colo.: Westview Press, 1983.

Kievitt, Frank David. "*A Canticle for Leibowitz* as a Third Testament." In *The Transcendent Adventure: Studies of Religion in Science Fiction/Fantasy*, ed. Robert Reilly. Westport, Conn.: Greenwood Press, 1985.

Klemm, David E. "The Rhetoric of Theological Argument." In *The Rhetoric of the Human Sciences: Language and Argument in Scholarship and Public Affairs*, ed. John S. Nelson, Allan Megill, and Donald N. McCloskey. Madison: University of Wisconsin Press, 1987.

Klotz, Irving M. "Postmodernist Rhetoric Does Not Change Fundamental Scientific Facts." *The Scientist* 10.15 (July 22, 1992): 9–12.

Krauss, Lawrence. "In Defense of Nonsense." Editorial. *New York Times*, July 30, 1996. Available at http://www.drizzle.com/jwalsh/sokal/articles/lkrauss.html.

Kuhn, Thomas S. *The Structure of Scientific Revolutions*. 2d ed. Chicago: University of Chicago Press, 1970.

Kurdas, Chidem. "Accumulation and Technical Change: Marx Revisited." *Science and Society* 59.1 (spring. 1995): 52.

Kurzman, Charles. "The Rhetoric of Science: Strategies for Logical Leaping." *Berkeley Journal of Sociology* 33 (1988): 131–58.

Lakoff, George, and Mark Johnson. *Metaphors We Live By*. Chicago: University of Chicago Press, 1980.

Lamont, Lansing. *Day of Trinity*. New York: Atheneum, 1965.

Landau, Misia. "Paradise Lost: The Theme of Terrestriality in Human Evolution." In *The Rhetoric of the Human Sciences: Language and Argument in Scholarship and Public Affairs*, ed. John S. Nelson, Allan Megill, and Donald N. McCloskey. Madison: University of Wisconsin Press, 1987.

Lapp, Ralph E. "The Hydrogen Bomb: IV." *Scientific American* 182.6 (June 1950): 13–15.

Leff, Michael C. "Modern Sophistic and the Unity of Rhetoric." *The Rhetoric of the Human Sciences: Language and Argument in Scholarship and Public Affairs*, ed. John S. Nelson, Allan Megill, and Donald N. McCloskey. Madison: University of Wisconsin Press, 1987.

Leontief, Wassily. "Machines and Man." *Scientific American* 187.3 (September 1952): 153–60.

Lie, Trygve. "UN v. Mass Destruction." *Scientific American* 182.1 (January 1950): 11–15.

Lockheed Aircraft Corporation. Advertisement. *Scientific American* 189.1 (July 1953): 85.

———. Advertisement. *Scientific American* 186.3 (March 1952): 37.

Lyne, John, and Henry F. Howe. "'Punctuated Equilibria': Rhetorical Dynamics of a Scientific Controversy." In *Landmark Essays on Rhetoric of Science*, ed. Randy Allen Harris. Mahwah, N.J.: Hermagoras Press, 1997.

Mann, Martin. "U.S. 'Space Fence' on Alert for Russian Spy Satellites." *Popular Science* 175.1 (July 1959): 62-198.

Massachusetts Mutual Life Insurance. Advertisement. *Time*, January 7, 1952, 43.

McGuire, J. E., and Trevor Melia. "The Rhetoric of the Radical Rhetoric of Science." *Rhetorica* 9.4 (fall 1991): 301-16.

———. "Some Cautionary Strictures on the Writing of the Rhetoric of Science." *Rhetorica* 7.1 (winter 1989): 206-15.

McMillen, Liz. "Scholars Who Study the Lab Say Their Work Has Been Distorted." *Chronicle of Higher Education*, June 28, 1996, A8.

Medhurst, Martin J., ed. *Cold War Rhetoric: Strategy, Metaphor, and Ideology*. New York: Greenwood Press, 1990.

———. "Eisenhower's 'Atoms for Peace' Speech: A Case Study in the Strategic Use of Language." *Cold War Rhetoric: Strategy, Metaphor, and Ideology*. New York: Greenwood Press, 1990.

Megill, Allan, and Donald N. McCloskey. "The Rhetoric of History." In *The Rhetoric of the Human Sciences: Language and Argument in Scholarship and Public Affairs*, ed. John S. Nelson, Allan Megill, and Donald N. McCloskey. Madison: University of Wisconsin Press, 1987.

Melpar, Inc. Advertisement. *Scientific American* 189.1 (July 1953): 46.

"Milestone." *Scientific American* 189.1 (July 1953): 40-41.

"Military Science." *Scientific American* 187.5 (November 1952): 36.

Miller, Walter M., Jr. *A Canticle for Leibowitz*. Philadelphia: J. B. Lippincott Co., 1959.

Moore Business Forms. Advertisement. *Time*, January 7, 1952, 52.

Moore, James. "Science and Religion: Some Historical Perspectives, By John Hedley Brooke." *History of Science* 30.89 (September 1, 1992): 311.

Morrissey, Thomas J. "Armageddon from Huxley to Hoban." *Extrapolation* 25 (1984): 197-213.

Moss, Norman. *Men Who Play God: The Story of the Hydrogen Bomb*. Baltimore: Penguin Books, 1972.

Nelkin, Dorothy. "The 'Science Wars': What Is at Stake?" *Chronicle of Higher Education*, July 26, 1996. Available online at http://www.drizzle.com/jwalsh/sokal/articles/dnelkin.html.

Nelson, John S. "Stories of Science and Politics: Some Rhetorics of Political Research." In *The Rhetoric of the Human Sciences: Language and Argument in Scholarship and Public Affairs*, ed. John S. Nelson, Allan Megill, and Donald N. McCloskey. Madison: University of Wisconsin Press, 1987.

Nelson, John S., Allan Megill, and Donald N. McCloskey. "Rhetoric of Inquiry." In *The Rhetoric of the Human Sciences: Language and Argument in Scholarship and Public Affairs,* ed. John S. Nelson, Allan Megill, and Donald N. McCloskey. Madison: University of Wisconsin Press, 1987.

O'Hear, Anthony. "Science and Religion." *British Journal for the Philosophy of Science* 44.3 (September 1, 1993): 505.

Parry-Giles, Shawn J. "Rhetorical Experimentation and the Cold War, 1947–1953: Development of an Internationalist Approach to Propaganda." *Quarterly Journal of Speech* 80.4 (November 1, 1994): 448.

Pollitt, Katha. "Pomolotov Cocktail." *The Nation,* June 10, 1996. Available online at http://www.thenation.com/issue/960610/0610poll.htm.

P. R. Mallory & Co. Advertisement. *Scientific American* 189.1 (July 1953): 7.

Prelli, Lawrence J. "The Rhetorical Construction of Scientific Ethos." In *Landmark Essays on Rhetoric of Science,* ed. Randy Allen Harris. Mahwah, N.J.: Hermagoras Press, 1997.

———. "Rhetorical Logic and the Integration of Rhetoric and Science." *Communication Monographs* 57.4 (December 1, 1990): 315–21.

———. *Rhetoric Review* 9.2 (spring 1991): 369–72.

Quine, W. V. *Ontological Relativity and Other Essays.* New York: Columbia University Press, 1969.

"Raw Materials for the U.S." *Scientific American* 188.4 (April 1953): 44.

Reeves, Carol. "The Rhetoric of Scientific Discovery Accounts." In *Landmark Essays on Rhetoric of Science,* ed. Randy Allen Harris. Mahwah, N.J.: Hermagoras Press, 1997.

Restivo, Sal. *Science, Society, and Values: Toward a Sociology of Objectivity.* Bethlehem: Lehigh University Press, 1994.

Ridenour, Louis N. "The Hydrogen Bomb." *Scientific American* 182.3 (March 1950): 11–15.

Robbins, Bruce. "On Being Hoaxed." Draft of article for *Tikkun,* circulated on the Internet in August 1996. Available online at http://www.drizzle.com/jwalsh/sokal/articles/rbbns2tkkn.html.

———. "Reality and Social Text." *In These Times,* July 8, 1996. Available online at http://www.drizzle.com/jwalsh/sokal/articles/rbbns2itt.html.

Robbins, Bruce, and Andrew Ross. [*Social Text*'s editorial response to Sokal, in "Mystery Science Theater"]. *Lingua franca* 6 (July/August 1996): 54–64. Available online at http://www.nyu.edu/pubs/socialtext/sokal.html.

Roe, Anne. "A Psychologist Examines 64 Eminent Scientists." *Scientific American* 187.5 (November 1952): 21–25.

Rorty, Richard. "Science as Solidarity." In *The Rhetoric of the Human Sciences: Language and Argument in Scholarship and Public Affairs,* ed. John S. Nelson, Allan Megill, and Donald N. McCloskey. Madison: University of Wisconsin Press, 1987.

Rosaldo, Renato. "Where Objectivity Lies: The Rhetoric of Anthropology." In *The Rhetoric of the Human Sciences: Language and Argument in Scholarship and Pub-*

lic Affairs, ed. John S. Nelson, Allan Megill, and Donald N. McCloskey. Madison: University of Wisconsin Press, 1987.

Rosen, Ruth. "A Physics Prof Drops a Bomb on the Faux Left." *Los Angeles Times,* May 23, 1996, A11.

Russell, Bertrand. *The Impact of Science on Society.* New York: Simon and Schuster, 1953.

Sagan, Carl. *The Demon-Haunted World: Science as a Candle in the Dark.* New York: Random House, 1996.

Schiappa, Edward. "The Rhetoric of Nukespeak." *Communication Monographs* 56 (1989): 251-72.

Schoen, E. L. "The Roles of Predictions in Science and Religion." *International Journal for Philosophy of Religion* 29.1 (February 1, 1991): 1.

"Science in the Budget." *Scientific American* 186.3 (March 1952): 34.

Segal, Robert. "Paralleling Religion and Science: The Project of Robert Horton." *Annals of Scholarship* 10.2 (1993): 177.

Segre, Eduardo. "Religion versus Science?" *American Journal of Physics* 62.4 (April 1, 1994): 296.

Senior, W. A. "'From the begetting of monster': Distortion as Unifier in *A Canticle for Leibowitz.*" *Extrapolation* 34.4 (1993): 329-39.

Serling, Rodman Edward. "Rod Serling: The Creator of *Twilight Zone.*" February 5, 1997. Available at http://www.scifi.com/twizone/twilite3.html.

Shaheen, Jack G., and Richard Taylor. "The Beginning of the End." In *Nuclear War Films,* ed. Jack G. Shaheen. Carbondale: Southern Illinois University Press, 1978.

Shapiro, Michael J. "The Rhetoric of Social Science: The Political Responsibilities of the Scholar." In *The Rhetoric of the Human Sciences: Language and Argument in Scholarship and Public Affairs,* ed. John S. Nelson, Allan Megill, and Donald N. McCloskey. Madison: University of Wisconsin Press, 1987.

Shippey, T. A. "The Cold War in Science Fiction, 1940-1960." In *Science Fiction: A Critical Guide,* ed. Patrick Parrinder. London: Longman, 1979.

Shute, Nevil. *On the Beach.* New York: William Morrow and Co., 1957.

Siebers, Tobin. *Cold War Criticism and the Politics of Skepticism.* New York: Oxford University Press, 1993.

Simons, Herbert W., and Trevor Melia, eds. *The Legacy of Kenneth Burke.* Madison: University of Wisconsin Press, 1989.

Siracusa, Joseph M., ed. *The American Diplomatic Revolution: A Documentary History of the Cold War, 1941-1947.* Port Washington, N.Y.: Kennikat Press, 1977.

Snyder, Alvin A. *Warriors of Disinformation: American Propaganda, Soviet Lies, and the Winning of the Cold War.* New York: Arcade Publishing, 1995.

Sokal, Alan. "A Physicist Experiments with Cultural Studies." *Lingua Franca* (May/June 1996): 62-64. New York University. Available at http://www.physics.nyu.edu/faculty/sokal/lingua_franca_v4/lingua_franca_v4.html.

———. "Sokal's Reply to *Social Text* Editorial." New York University. September 23, 1996. Available at http://www.physics.nyu.edu/faculty/sokal/reply.html.

———. "Transgressing the Boundaries: Toward a Transformative Hermeneutics of Quantum Gravity." *Social Text* 14.1 (spring/summer 1996). Available online at http://www.physics.nyu.edu/faculty/sokal/transgress_v2/ transgress_v2_singlefile.html.

Sontag, Susan. "The Imagination of Disaster." *Against Interpretation*. New York: Noonday Press, 1966.

"Space Flight." *Scientific American* 186.5 (May 1952): 38.

Spector, Judith A. "Walter Miller's *A Canticle for Leibowitz*: A Parable for Our Time?" *Midwest Quarterly* 22 (summer 1981): 337–45.

Spencer, Susan. "The Post-Apocalyptic Library: Oral and Literate Culture in *Fahrenheit 451* and *A Canticle for Leibowitz*." *Extrapolation* 32.4 (1991): 331–42.

"The Steel-Hungry Nation Gets a Mighty New Mill." *Life*, December 22, 1952, 47–55.

Switzer, John. "Unofficial Summary of the Rush Limbaugh Show for Wednesday, May 22, 1996." Available online at http://www.drizzle.com/jwalsh/sokal/ articles/rlimbaugh.html.

Tenner, Edward. *Why Things Bite Back: Technology and the Revenge of Unintended Consequences*. New York: Alfred A. Knopf, 1996.

Tietge, David J. "The G.I. Bill." *War and American Popular Culture*. Westport, Conn.: Greenwood Press, 1999.

———. "The Role of Burke's Four Master Tropes in Scientific Expression." *Journal of Technical Writing and Communication* 28.4 (1998): 317–24.

Trout, B. Thomas. "Rhetoric Revisited: Political Legitimation and the Cold War." *International Studies Quarterly* 19 (September 1979): 251–84.

Ulam, Stanislaw M. *Science, Computers, and People: From the Tree of Mathematics*. Ed. Mark C. Reynolds and Gian-Carlo Rota. Boston: Birkhauser, 1986.

Union Carbide and Carbon Corporation. Advertisement. *Scientific American* 188.3 (March 1953): 72.

USDA. "Farmers and Land." http://www.USDA.gov.

Van Tassel, David D., and Michael G. Hall, eds. *Science and Society in the United States*. Homewood, Ill.: Dorsey Press, 1966.

Waddell, Craig. "The Role of Pathos in the Decision-Making Process: A Study in the Rhetoric of Science Policy." In *Landmark Essays on Rhetoric of Science*, ed. Randy Allen Harris. Mahwah, N.J.: Hermagoras Press, 1997.

Weart, Spencer R. *Nuclear Fear: A History of Images*. Cambridge, Mass.: Harvard University Press, 1988.

Weaver, Richard M. "Dialectic and Rhetoric at Dayton, Tennessee." In *Landmark Essays on Rhetoric of Science*, ed. Randy Allen Harris. Mahwah, N.J.: Hermagoras Press, 1997.

Werner, Vivian. *Scientist versus Society: Six Profiles*. New York: Hawthorn Books, 1975.

Wertheim, Margaret. "Science and Religion." *Omni* 17.1 (October 1, 1994): 36.

Westinghouse Electric Corporation. Advertisement. *Scientific American* 188.3 (March 1953): 62.

"Wheels Are Its Basic Tools." *Life,* December 15, 1952.

"Where the Money Goes." *Scientific American* 189.2 (August 1953): 40.

Wiebe, Donald. "Is Science Really an Implicit Religion?" *Studies in Religion* 18.2 (1989): 171.

———. "Religion, Science, and the Transformation of Knowledge." *Sophia* 32.2 (1993): 36.

Willems, Jaap. "The Biologist as a Source of Information to the Press." *Bulletin of Science, Technology, and Society* 15.1 (1995): 21.

Williams, Raymond. "Base and Superstructure in Marxist Cultural Theory." *Contemporary Literary Criticism.* 2d ed. New York: Blackwell, 1993.

Wilson, Edward O. *Consilience: The Unity of Knowledge.* New York: Vintage Books, 1998.

Yahraes, Herbert. "The Tragic Truth about Taking Dope." *Popular Science* 175.4 (October 1959): 81-260.

York, Herbert F. *The Advisors: Oppenheimer, Teller, and the Superbomb.* San Francisco: W. H. Freeman and Co., 1976.

Young, Marilyn. "The Rhetorical Origins of American Anti-Communism." Pittsburgh, Pa.: University of Pittsburgh Press, 1970.

Index